SPRINGER PROCEEDINGS IN PHYSICS 105

SPRINGER PROCEEDINGS IN PHYSICS

87 **Proceedings of the 25th International Conference on the Physics of Semiconductors**
Editors: N. Miura and T. Ando

88 **Starburst Galaxies**
Near and Far
Editors: L. Tacconi and D. Lutz

89 **Computer Simulation Studies in Condensed-Matter Physics XIV**
Editors: D.P. Landau, S.P. Lewis, and H.-B. Schüttler

90 **Computer Simulation Studies in Condensed-Matter Physics XV**
Editors: D.P. Landau, S.P. Lewis, and H.-B. Schüttler

91 **The Dense Interstellar Medium in Galaxies**
Editors: S. Pfalzner, C. Kramer, C. Straubmeier, and A. Heithausen

92 **Beyond the Standard Model 2003**
Editor: H.V. Klapdor-Kleingrothaus

93 **ISSMGE**
Experimental Studies
Editor: T. Schanz

94 **ISSMGE**
Numerical and Theoretical Approaches
Editor: T. Schanz

95 **Computer Simulation Studies in Condensed-Matter Physics XVI**
Editors: D.P. Landau, S.P. Lewis, and H.-B. Schüttler

96 **Electromagnetics in a Complex World**
Editors: I.M. Pinto, V. Galdi, and L.B. Felsen

97 **Fields, Networks, Computational Methods and Systems in Modern Electrodynamics**
A Tribute to Leopold B. Felsen
Editors: P. Russer and M. Mongiardo

98 **Particle Physics and the Universe**
Proceedings of the 9th Adriatic Meeting, Sept. 2003, Dubrovnik
Editors: J. Trampetić and J. Wess

99 **Cosmic Explosions**
On the 10th Anniversary of SN1993J (IAU Colloquium 192)
Editors: J. M. Marcaide and K. W. Weiler

100 **Lasers in the Conservation of Artworks**
LACONA V Proceedings, Osnabrück, Germany, Sept. 15–18, 2003
Editors: K. Dickmann, C. Fotakis, and J.F. Asmus

101 **Progress in Turbulence**
Editors: J. Peinke, A. Kittel, S. Barth, and M. Oberlack

102 **Adaptive Optics for Industry and Medicine**
Proceedings of the 4th International Workshop
Editor: U. Wittrock

103 **Computer Simulation Studies in Condensed-Matter Physics XVII**
Editors: D.P. Landau, S.P. Lewis, and H.-B. Schüttler

104 **Complex Computing-Networks**
Brain-like and Wave-oriented Electrodynamic Algorithms
Editors: I.C. Göknar and L. Sevgi

105 **Computer Simulation Studies in Condensed-Matter Physics XVIII**
Editors: D.P. Landau, S.P. Lewis, and H.-B. Schüttler

106 **Modern Trends in Geomechanics**
Editors: W. Wu and H.S. Yu

107 **Microscopy of Semiconducting Materials**
Proceedings of the 14th Conference, April 11–14, 2005, Oxford, UK
Editors: A.G. Cullis and J.L. Hutchison

Volumes 60–86 are listed at the end of the book.

D.P. Landau S.P. Lewis H.-B. Schüttler
(Eds.)

Computer Simulation Studies in Condensed-Matter Physics XVIII

Proceedings of the Eighteenth Workshop
Athens, GA, USA, March 7–11, 2005

With 72 Figures

Springer

Professor David P. Landau, Ph.D.
Professor Steven P. Lewis, Ph.D.
Professor Heinz-Bernd Schüttler, Ph.D.
Center for Simulational Physics
The University of Georgia
Athens, GA 30602-2451, USA

ISSN 0930-8989

ISBN:13 978-3-642-06909-3 e-ISBN:13 978-3-540-32640-3

Bibliographic information published by Die Deutsche Bibliothek
Die Deutsche Bibliothek lists this publication in the Deutsche Nationalbibliografie; detailed bibliographic data is available in the Internet at <http://dnb.ddb.de>.

Springer is a part of Springer Science+Business Media.

springer.com

Softcover reprint of the hardcover 1st edition 2006

Cover concept: eStudio Calamar Steinen
Cover production: WXMDesign GmbH, Heidelberg

Preface

Almost two decades ago, because of the tremendous increase in the power and utility of computer simulations, The University of Georgia formed the first institutional unit devoted to the use of simulations in research and teaching: The Center for Simulational Physics. As the international simulations community expanded further, we sensed a need for a meeting place for both experienced simulators and neophytes to discuss new techniques and recent results in an environment which promoted lively discussion. As a consequence, the Center for Simulational Physics established an annual workshop on Recent Developments in Computer Simulation Studies in Condensed Matter Physics. This year's workshop was the eighteenth in this series, and the continued interest shown by the scientific community demonstrates quite clearly the useful purpose that these meetings have served. The latest workshop was held at The University of Georgia, March 7–11, 2005, and these proceedings provide a "status report" on a number of important topics. This volume is published with the goal of timely dissemination of the material to a wider audience.

We wish to offer a special thanks to IBM for partial support of this year's workshop.

This volume contains both invited papers and contributed presentations on problems in both classical and quantum condensed matter physics. We hope that each reader will benefit from specialized results as well as profit from exposure to new algorithms, methods of analysis, and conceptual developments.

Athens, GA, USA
October 2005

D.P. Landau
S.P. Lewis
H.-B. Schüttler

Preface

Contents

1

Computer Simulation Studies in Condensed Matter Physics: An Introduction

D.P. Landau, S.P. Lewis, and and H.-B. Schüttler

Center for Simulational Physics, The University of Georgia,
Athens, GA 30602-2451, USA

Computer simulation studies in condensed matter physics now play a fundamental role in many areas of investigation. The "status report" which is contained in this volume is the result of presentations and discussion that occurred during the 18th Annual Workshop at the Center for Simulational Physics. The texts of both longer, invited presentations as well as a number of contributed papers are included. The reader will find that the scope of simulational/computational studies is broad and that substantial potential for cross-fertilization of methods between different sub-fields is evident.

The volume opens with four papers on quantum simulations and quantum computing. First, Matsumoto and Takayama report on impurity-enhanced phase transitions in low-dimensional quantum magnets, described by a quasi-one-dimensional, $S = 1$ Heisenberg antiferromagnet host lattice with weak interchain coupling and spin $S = 1/2$ or $S = 0$ impurities. Monte Carlo results for the impurity-enhanced Néel transition temperature are interpreted in terms of a "valence bond solid" picture for the pure host system. Next, Harada generalizes the world-line loop algorithm and applies it to certain highly symmetric quantum spin models with spin quantum numbers $S > 1/2$. The algorithm uses a representation, and updates, of spin world-lines in terms of so-called non-binary loops. Results for the suppression of long-range order and the emergence of a short-range RVB state in two-dimensional $SU(N)$ quantum anti-ferromagnets at large N are discussed. De Raedt, De Raedt, and Michielsen then describe an entirely classical, deterministic, event-based model processing unit, having only very primitive learning capabilities, which can simulate the probability distributions of a corresponding quantum mechanical system. A classical simulation performed by such a "deterministic learning machine" (DLM) reproduces the photon count distribution of an idealized single-photon Mach–Zehnder interferometer. In a companion paper, Michielsen, De Raedt and De Raedt discuss how their proposed DLM unit, as well as a stochastically modified version, can be used to build classical processing networks that will simulate quantum computing networks. They

Springer Proceedings in Physics, Volume 105
Computer Simulation Studies in Condensed-Matter Physics XVIII
Eds.: D.P. Landau, S.P. Lewis and H.-B. Schüttler

illustrate this by a DLM network which can reproduce the results of a corresponding quantum computing network executing a simple application of Shor's factorization algorithm.

In Part II, materials and nanostructures are highlighted. In the first of three invited presentations, Abraham uses classical atomistic dynamics to study crack-motion instabilities in brittle fracture, in order to shed light on the mismatch between theory and experiment on the conditions that give rise to instabilities. His results reveal the dominant role played by hyperelasticity, the nonlinear elasticity relevant to crystal deformation, in the instability dynamics. Moreover, he describes a universal scaling law for the instability phenomenon based on an effective hyperelasticity modulus. Kronik then applies state-of-the-art first-principles electronic structure methods to explore two exciting recent developments in semiconductors research: semiconductor nanocrystals and dilute magnetic semiconductors. In particular, he studies quantum size effects in Si and Ge nanocrystals and examines the spin-polarized electronic properties of the dilute magnetic semiconductors, MnGaAs and MnGaN. Finally he weds the two major developments by studying prototype dilute magnetic semiconducting nanocrystals. In the next paper Hadjisavvas and Kelires conduct Monte Carlo simulations to study interfacial properties and strain and ordering processes in nanostructured materials. Their work reveals the importance of internal stress fields in establishing the composition profile of Ge islands on Si (100). Furthermore they study the size-dependence of interfacial structure, stability, and disorder for Si nanocrystals embedded in a-SiO_2. Lastly, Puzyrev and Faulkner carry out molecular dynamics simulations using embedded-atom potentials to determine equilibrium atomic displacements in the Cu–Au alloy family. Their goal is to elucidate the role of atomic size effects in alloy structure and to make contact with scattering experiments.

Part III contains seven papers that examine diverse questions in statistical mechanics. In the first of these, Tomita et al. use heat bath Monte Carlo simulations to examine possible ground states for dipolar systems on several two and three dimensional lattices. For triangular and bcc lattices, magnetic domain states turn out to be stable. Then Boettcher explores low temperature states in Ising spin glasses at the percolation threshold. The finite size behavior of the defect energy is used to extract the critical exponent at percolation. Oguchi, Yukawa and Ito then report on a non-equilibrium molecular dynamics study of heat conduction of a Lennard–Jones system. The density profile of the liquid-gas interface is determined and compared with theoretical predictions. In the following paper Georgiev, Schmittmann, and Zia perform a Monte Carlo simulation of a two-lane driven diffusive system. They discuss the dynamic scaling of coarsening particle clusters and find indications of a dynamic precursor to changes in the steady states as a parameter is varied. Hurt et al. then describe a Wang–Landau sampling study of the Blume–Capel model on a square lattice. The tricritical point is located and the effectiveness of the algorithm is discussed. The time dependence of the Contact Process is discussed next by Ramasco et al. As a control parameter is varied they find a critical point and an indication of a violation of the fluctuation-dissipation theorem. This part concludes with a study of the effects of size dispersity on two-dimensional melting by Watanabe, Yukawa, and Ito. From the non-equilibrium behavior of the bond-orientational order parameter they conclude that the dispersity may not only destroy order but, under certain conditions, may also give rise to new structures.

New algorithmic developments are featured in Part IV, beginning with a proposal by Pruessner and Moloney for a novel asynchronous distributed parallelization approach for Monte Carlo simulations of percolation, based on the Hoshen–Kopelman algorithm. High precision studies of very large lattices are illustrated by calculations of several distributions of model observables in a variety of two-dimensional percolation model topologies. In the next paper Adler and Rosen present a brief overview of AViz, a recently developed instructional visualization system for atomistic simulations and visualizations in condensed matter physics. They give examples of projects, suitable for senior-level undergraduates, including atomistic, spin, polymer and lattice simulations and AViz based visualizations. Shim and Amar then describe a study of performance characteristics of a recently developed semi-rigorous synchronous algorithm for parallel kinetic Monte Carlo simulations, applied to simple epitaxial growth models. Results for the parallel efficiency, accuracy, and scalability as a function of the relevant growth model parameters are presented. To close this part, Wu, et al. present results for the performance of the flat-histogram Monte Carlo algorithm applied to Ising–Potts models in varying dimensions. They study the scaling of the "mean first passage time to span the entire energy space" as a function of system size and dimensionality. Improvements of the algorithm, by way of cluster dynamics and ensemble optimization techniques, are proposed.

The last part of this volume contains three papers on biological and soft condensed matter. Opening this Part is an invited paper by Schilling, Vink, and Wolfsheimer, in which they perform grand canonical Monte Carlo simulations to study the isotropic-to-nematic phase transition in anisotropic colloid suspensions, using soft spherocylinders to model the colloidal particles. Their focus is on determining properties of phase coexistence, especially interface tension, in order to make contact with existing theories. Next, Guo and Luijten apply Langevin dynamics to simulate the translocation of a polymer through a nanopore under various conditions. They show that the scaling exponent relating translocation time to polymer chain length depends on the strength of the driving force, with a crossover from equilibrium to non-equilibrium behavior for the polymer during translocation for increasing driving force. To close this volume, Yasi, Korniss, and Caraco studied the spread of an advantageous mutant in a two-allele population, using dynamic Monte Carlo simulation on a simple lattice-based population dynamics model. This model exhibits homogeneous nucleation behavior for the spread of advantageous mutation.

Part I

Quantum Simulations and Quantum Computing Nanostructures

2

Quantum Monte Carlo Simulations of the Impurity-Induced Phase Transitions in Low-Dimensional Magnets

M. Matsumoto[1] and H. Takayama[2]

[1] Department of Physics, Graduate School of Science, Tohoku University, Sendai 980-8578, Japan
[2] Institute for Solid State Physics, University of Tokyo, Chiba 277-8581, Japan

Abstract. We study the impurity-induced phase transitions in a quasi-one-dimensional Heisenberg antiferromagnet doped with magnetic spin-1/2 impurities and non-magnetic ones. The impurity-induced transition temperature determined by the quantum Monte Carlo method with the continuous-time loop algorithm is monotonically increasing as a function of the magnitude of the impurity spin. To these results, we give discussions based on the valence-bond solid-like picture for the pure system and the inspection of the local magnetic structure around the impurities.

2.1 Motivation

Impurities in the low-dimensional quantum magnets drastically change the magnetic properties of the host materials. For the low-dimensional magnets such as $CuGeO_3$ [1] and $PbNi_2V_2O_8$ [2], pure systems have disordered ground states, while non-magnetic-impurity-doped ground states are antiferromagnetically ordered. These impurity-induced ordered states have been given a qualitative understanding based on the valence-bond solid (VBS) picture [3] for low-dimensional magnets. According to the VBS picture, a spin with the magnitude S on each site is treated as ferromagnetically coupled $2S$ subspins with the magnitude 1/2, and the disordered ground state of a low-dimensional antiferromagnet can be seen as a state made of closely packed spin-1/2 singlet pairs, which are called valence bonds. When a non-magnetic impurity is doped, a valence bond is broken after one of the spin in a pair is lost and the counterpart subspin is liberated. Thus non-magnetic impurities induce nearly-free local magnetic moments around their doped sites and there is long-range correlation between them that keeps the bulk antiferromagnetic long-range order [4].

Recently, in the experiments using the spin-1 gapped magnet $PbNi_2V_2O_8$, the effects of several species of magnetic impurities were systematically investigated and it was found that the impurity-induced transition temperatures

Springer Proceedings in Physics, Volume 105
Computer Simulation Studies in Condensed-Matter Physics XVIII
Eds.: D.P. Landau, S.P. Lewis and H.-B. Schüttler

show non-monotonic dependence on the magnitude of the impurity spin [5]. In particular, compared with the transition temperatures induced by non-magnetic impurities, those done by Cu^{2+} are found to be much lower. These results can be thought to be reasonable if we inspect the magnitude of the local magnetic moments that are expected to appear around impurities by the simplest picture analogous to the VBS one. There are two spin-1/2 effective magnetic moments around one non-magnetic impurity, and these are coupled by an effective ferromagnetic coupling mediated by the interchain couplings, making one spin-1 local magnetic moment. For a magnetic spin-1/2 impurity, there is one spin-1/2 local magnetic moment per one impurity. Thus the magnitude of the local magnetic moments is larger for the non-magnetic impurities than for the magnetic spin-1/2 ones. We can at least hope to understand the experimental result that the spin-1/2 impurities induce much lower transition temparature than the non-magnetic ones do. Motivated by this observation, we do quantum Monte Carlo simulations for the quasi-one-dimensional Heisenberg antiferromagnets with site impurities and see if our picture for the impurity-induced ordered state is valid or not.

2.2 Model and Method

We introduce our model following the observation described in the previous section. The pure system is weakly coupled spin-1 antiferromagnetic Heisenberg chains aligned in parallel on a simple cubic lattice. The Hamiltonian is written as follows

$$\begin{aligned} \mathcal{H} = J \sum_{x,y,z} \boldsymbol{S}_{x,y,z} \cdot \boldsymbol{S}_{x+1,y,z} \\ + J' \left(\sum_{x,y,z} \boldsymbol{S}_{x,y,z} \cdot \boldsymbol{S}_{x,y+1,z} + \sum_{x,y,z} \boldsymbol{S}_{x,y,z} \cdot \boldsymbol{S}_{x,y,z+1} \right) . \end{aligned} \tag{2.1}$$

Here the spin operator $\boldsymbol{S}_{x,y,z}$ has the magnitude $S \equiv |\boldsymbol{S}| = 1$ and x, y, z denote the points on a simple cubic lattice. We consider only the nearest neighbor antiferromagnetic couplings. The intrachain (interchain) coupling is denoted by $J(J') > 0$. We set the x axis parallel to the chains. The strength of the interchain coupling, J', is set to be $0.01J$. This value is small enoungh to allow the Haldane gap [6] to be finite even in a three-dimensional space. The critical point between the gapped phase and the antiferromagnetically ordered one, J'_{c}, is estimated to be $J'_{\mathrm{c}} = 0.013J$[3] [8].

In the impurity-doped system, the $S = 1$ host spins are randomly replaced with the $S \neq 1$ impurity spins with the couplings between the spins left

[3] Recently the Néel temperatures of quasi-low-dimensional antiferromagnets were investigated systematically and modified random phase approximation theory was discussed, in which the position of the critical point between the gapped phase and the long-range ordered phase can be estimated; see [7].

unchanged. The aim of using this simple model is to let the magnitude of local magnetic moments play a dominant role in determining the impurity-induced transition temperatures by letting all of the species of impurities share the equal strength of the effective couplings between the impurity neighborhoods when the concentration of the impurities is fixed. By this model we try to simulate the impurity-induced transition in which the non-magnetic impurities yield higher transition temperature than the magnetic spin-1/2 ones do.

The impurity-induced transition temperatures are found as follows. The physical observables are calculated by the quantum Monte Carlo with the continuous-time loop algorithm [9, 10]. The correlation length is calculated using the second moment of the two-point correlation function [10, 11]. The ratio of the correlation length to the linear size of the system is plotted with respect to the temperature for several system sizes and the crossing point (if any is found) gives the position of critical point. By this simple analysis, enough accuracy in the data for the qualitative discussion in the present study is obtained. The concentration of impurities is fixed to be 10% and the impurity-induced transition temperatures (if any is found) are compared with each other between the species of impurities.

2.3 Results and Discussions

By the procedure described in the previous section, we find that all species of impurities induce phase transitions in our model. Specifically, it is found that non-magnetic impurities induce the Néel temperature at $T_{\rm N} = 0.03 \pm 0.01$ and magnetic spin-1/2 ones do that at $T_{\rm N} = 0.05 \pm 0.01$. As for the other magnetic-impurity-induced transition temperatures, we find that they are monotonically increasing as a function of the magnitude of the spin of impurities [12]. We must note that the transition temperature induced by the magnetic spin-1/2 impurities is higher than that done by the non-magnetic ones, which is in contrast with our expectation before simulations.

To check our presumption that the magnitude of the local magnetic moments, which should be relevant in the value of the impurity-induced transition temperature, must be larger for non-magnetic impurities than for magnetic spin-1/2 ones, we are going to take a look directly on the local magnetic moments around an impurity. We calculate the local field susceptilities for each site i, which we denote by $\chi_{{\rm local},i}$, on a single spin-1 chain with one impurity doped in the center site of the chain. The local field susceptibility is defined by the following formula

$$\chi_{{\rm local},i} = \left.\frac{\partial \langle S_i^z \rangle}{\partial h_i}\right|_{h_i=0} = \int_0^\beta d\tau \, \langle S_i^z(\tau) S_i^z(0) \rangle \, . \tag{2.2}$$

Here h_i is the local magnetic field applied on the site i, τ is the imaginary time introduced in the path-integral formalism that is employed in the loop

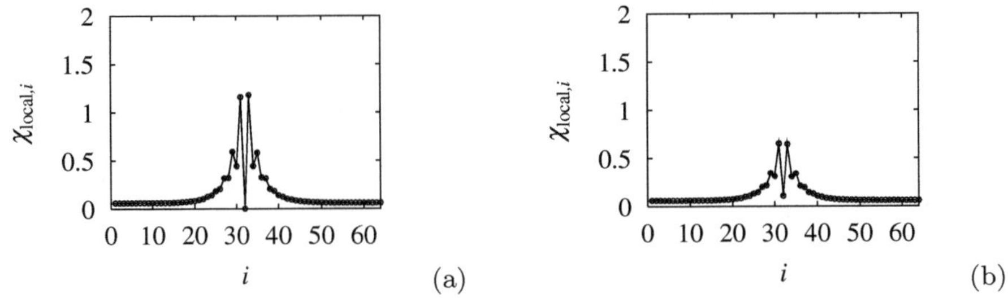

Fig. 2.1. Local field susceptibilities plotted with respect to the site indices of a spin-1 Haldane chain with an impurity doped in the center of it. The length of the chain is 64 and the temperature is 0.01 measured by the strength of the superexchange coupling. An impurity is (**a**) non-magnetic and (**b**) magnetic with spin magnitude 1/2

algorithm, and β is the inverse temperature. The real-space distribution of the local field susceptibility is shown in Fig. 2.1. Here the length of the chain is 64 and the temperature is $0.01J$ at which the pure system is well near the ground state. We see that the total magnitude of the local magnetic moments is larger in the neighborhood of a non-magnetic impurity than in that of a magnetic spin-1/2 one. This result is consistent with our first expectation.

As the transition temperature is expected to be determined by both of the magnitude of the magnetic moments that contribute to the long-range order and the strength of the interaction between them, we are led to consider the interaction between the local magnetic moments more closely. First we consider the neighborhoods of non-magnetic impurities. The two local magnetic moments with the effective magnitude 1/2 around a non-magnetic impurity are ferromagnetically coupled mediated by the very weak interchain couplings. The strength of this coupling is estimated to be of the order J'^2. Near the transition temperature that is in energy scale comparable to the strength of the interchain coupling J', the coupling of the order J'^2 between the two moments hardly contribute to the global antiferromagnetic order of the present interest. On the other hand, around a magnetic spin-1/2 impurity, two effective spin-1/2 moments at the edge of the pure chain and the impurity spin are directly coupled yielding an effective spin of the magnitude 1/2. See Fig. 2.2. When the ground state of the three spins with the magnitude 1/2 which are separated from the rest is inspected, they exhibit magnetic moments (whose magnitudes are less than 1/2) antiferromagnetically aligned. This strongly suggests that the three spins contribute to the global antiferromagnetic order even at temperatures of the order J', i.e., near the transition temperature. Thus it is reasonable to encounter the present simulational results that the impurity-induced temperatures are higher for the magnetic impurities than for the non-magnetic ones. Our first expectation may hold good in the ordering at temperatures below the order J'^2.

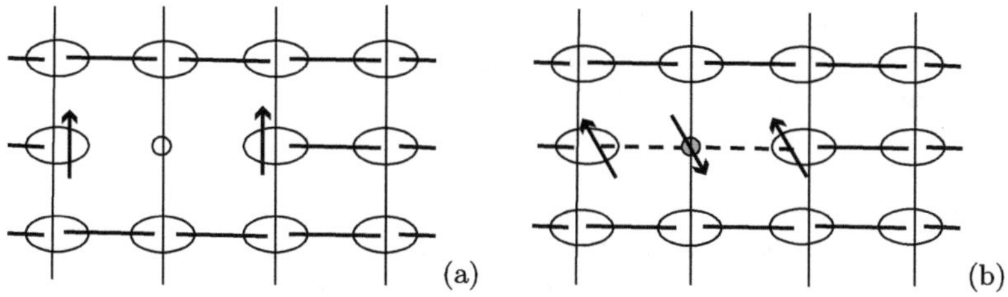

Fig. 2.2. Schematic figure for the local spin-spin couplings between local magnetic moments around (**a**) a non-magnetic impurity and (**b**) a spin-1/2 impurity in the gapped phase of the quasi-one-dimensional spin-1 Heisenberg antiferromagnet. The bold solid line represents a valence bond and the dotted line do the intrachain coupling. The arrows denote local spin-1/2 effective magnetic moments

In real experiments, various kinds of interaction around impurities must be playing key roles, which may require the detailed examination specific to each kind of impurities [5].

Acknowledgements

One of the authors (MM) thanks Prof. T. Masuda and Prof. K. Uchinokura for their useful comments and Dr. C. Yasuda and Dr. S. Todo for helpful discussions. The computations for this work were done on the SGI ORIGIN 2800 that was at the Supercomputer center in the Institute for Solid State Physics, University of Tokyo until this February. The codes are based on the library "LOOPER version 2" developed by Dr. S. Todo and Dr. K. Kato, and "PARAPACK version 2" developed by Dr. S. Todo.

References

1. M. Hase, I. Terasaki, K. Uchinokura: Phys. Rev. Lett. **70**, 3651 (1993); M. Hase, I. Terasaki, Y. Sasago, K. Uchinokura, H. Obara: Phys. Rev. Lett. **71**, 4059 (1993)
2. Y. Uchiyama, Y. Sasago, I. Tsukada, K. Uchinokura, A. Zheludev, T. Hayashi, N. Miura, P. Böni: Phys. Rev. Lett. **83**, 632 (1999)
3. I. Affleck, T. Kennedy, E.H. Lieb, H. Tasaki: Phys. Rev. Lett. **59**, 799 (1987)
4. C. Yasuda, S. Todo, M. Matsumoto, H. Takayama: Phys. Rev. B **64**, 092405 (2001)
5. S. Imai, T. Masuda, T. Matsuoka, and K. Uchinokura: preprint (cond-mat/0402595)
6. F.D.M. Haldane: Phys. Lett. **A93**, 464 (1983); Phys. Rev. Lett. **50**, 1153 (1983)
7. C. Yasuda, S. Todo, K. Hukushima, F. Alet, M. Keller, M. Troyer, H. Takayama: preprint (cond-mat/0312392)
8. T. Sakai, M. Takahashi: J. Phys. Soc. Jpn. **58**, 3131 (1989)

9. H.G. Evertz, G. Lana, M. Marcu: Phys. Rev. Lett. **70**, 875 (1993); B.B. Beard, U.-J. Wiese: Phys. Rev. Lett. **77**, 5130 (1996); excellent reviews on the loop algorithm are given in the following articles: H.G. Evertz: Adv. Phys. **52**, 1 (2003); N. Kawashima, K. Harada: J. Phys. Soc. Jpn. **73**, 1379 (2004); see also the article by K. Harada in this volume.
10. S. Todo, K. Kato: Phys. Rev. Lett. **87**, 047203 (2001)
11. F. Cooper, B. Freedman, D. Preston: Nucl. Phys. **B210[FS6]**, 210 (1982)
12. M. Matsumoto, H. Takayama: preprint (cond-mat/0412676)

3

Quantum World-line Monte Carlo Method with Non-binary Loops and Its Application

K. Harada

Graduate School of Informatics, Kyoto University, Kyoto 606-8501, Japan

Abstract. A quantum world-line Monte Carlo method for high-symmetrical quantum models is proposed. Firstly, based on a representation of a partition function using the Matsubara formula, the principle of quantum world-line Monte Carlo methods is briefly outlined and a new algorithm using non-binary loops is given for quantum models with high symmetry as SU(N). The algorithm is called non-binary loop algorithm because of non-binary loop updatings. Secondary, one example of our numerical studies using the non-binary loop updating is shown. It is the problem of the ground state of two-dimensional SU(N) anti-ferromagnets. Our numerical study confirms that the ground state in the small $N(\leq 4)$ case is a magnetic ordered Neel state, but the one in the large $N(\geq 5)$ case has no magnetic order, and it becomes a dimer state.

3.1 Introduction

The quantum world-line Monte Carlo (QMC) method has a long history. It has been used in many numerical studies of condensed-matter physics since it first proposed in 1970s [1]. However QMC simulations suffered as the result of some defects in conventional QMC algorithms. For example, near a critical point, or at low temperatures, the correlation time in samples became extremely long, thus we could not calculate the canonical ensemble average with accuracy. There were same problems in the Monte Carlo simulation of a classical system. Fortunately, a new algorithm was proposed by Swendsen and Wang, which solves such problems in classical systems. It is called cluster algorithm. The idea of the cluster algorithm can apply to quantum cases. Indeed, Evertz, Lana and Marcu first proposed a new QMC algorithm based on the cluster algorithm, which called loop algorithm [2]. The loop algorithm evolves into recent powerful QMC algorithms (see reviews [3] and [4]).

In the present article, we will focus to the loop algorithm and the generalization for high-symmetrical quantum models. In Sect. 3.2, a useful representations of a partition function for recent QMC algorithms will be derived. In Sect. 3.3, the principle of QMC method and the detail of the loop algorithm

Springer Proceedings in Physics, Volume 105
Computer Simulation Studies in Condensed-Matter Physics XVIII
Eds.: D.P. Landau, S.P. Lewis and H.-B. Schüttler

will be reviewed. In Sect. 3.4, our generalization of the loop algorithm to a non-binary loop updating will be proposed. In Sect. 3.5, we will show one example of our numerical studies with the non-binary loop updating: the problem of the ground state of the two-dimensional SU(N) quantum anti-ferromagnets.

3.2 World-line Representations Based on the Matsubara Formula

A density operator has all informations of a quantum system in an environment. In particular, if the system is in the heat-bath whose inverse temperature is β, the density operator $\rho(\beta)$ is the exponential operator of the system Hamiltonian as $\rho(\beta) = \exp(-\beta\mathcal{H})$, where $\mathcal{H}$ is the Hamiltonian of a quantum system.

The Hamiltonian usually consists of two parts $\mathcal{H}_0$ and V. Here $\mathcal{H}_0$ and V denote diagonal and non-diagonal operators, respectively. The V generally represents a perturbation part of the Hamiltonian. Then the non-trivial part of the density operator, $\rho'(\beta)$, satisfies a special Bloch equation in which there is explicitly only the V operator:

$$\frac{\mathrm{d}\rho'(\beta)}{\mathrm{d}\beta} = -V(\beta)\rho'(\beta) \quad \left(\rho(\beta) \equiv \mathrm{e}^{-\beta(\mathcal{H}_0+V)} \equiv \mathrm{e}^{-\beta\mathcal{H}_0}\rho'(\beta)\right), \tag{3.1}$$

where $V(t)$ denotes the interaction picture of V operator as $V(t) \equiv \mathrm{e}^{t\mathcal{H}_0}V\mathrm{e}^{-t\mathcal{H}_0}$. The solution of (3.1) formally satisfies an integral equation as

$$\rho'(\beta) = I - \int_0^\beta \mathrm{d}t V(t)\rho'(t), \tag{3.2}$$

where an integral variable t is called *imaginary time*, because the Bloch equation is related to the Schrödinger equation through analytic continuation.

The solution of (3.2) is written as a series of multiple integrals of the product of interaction pictures of V:

$$\rho'(\beta) = I - \int_0^\beta \mathrm{d}t_1 V(t_1) + \int_0^\beta \mathrm{d}t_2 \int_0^{t_2} \mathrm{d}t_1 V(t_2)V(t_1) - \cdots. \tag{3.3}$$

This representation of a density operator is known as the Matsubara formula. The order of interaction pictures $V(t)$ is in descending order. Since the perturbation part V is usually the sum of local interaction Hamiltonians V_b, the density operator is a series of multiple integrals of the product of interaction pictures of V_b:

$$\rho(\beta) = \mathrm{e}^{-\beta\mathcal{H}_0} \sum_{n=0}^{\infty} \sum_{b_n,\cdots,b_1} \int_{\beta \geq t_n \geq \cdots \geq t_1 \geq 0} \mathrm{d}t_n \cdots \mathrm{d}t_1 (-1)^n V_{b_n}(t_n) \cdots V_{b_1}(t_1), \tag{3.4}$$

where $V = \sum_b V_b$.

The matrix element of operator AB is equal to the summation of products of each matrix elements of A and B.

$$\langle \Psi_A | AB | \Psi_B \rangle = \sum_{\{|\psi\rangle\}:\text{basis}} \langle \Psi_A | A | \psi \rangle \langle \psi | B | \Psi_B \rangle \,, \tag{3.5}$$

where the set $\{|\psi\rangle\}$ is an orthonormal basis in the quantum state space. Therefore, inserting an orthonormal basis $\{|\psi\rangle\}$ between two operators of integrand in the Matsubara formula, the partition function Z can be represented as a multiple integral with respect to three kinds of variables, ψ_i, b_i and t_i:

$$Z \equiv \mathrm{Tr}\rho(\beta) \tag{3.6}$$

$$= \sum_{n=0}^{\infty} \sum_{\psi_n,\cdots,\psi_1(\psi_{n+1}=\psi_1)} \sum_{b_n,\cdots,b_1} \int_{\beta \geq t_n \geq \cdots \geq t_1 \geq 0} W_n(\{\psi_i\},\{b_i\},\{t_i\}) \,, \tag{3.7}$$

$$W_n(\{\psi_i\},\{b_i\},\{t_i\}) \equiv \mathrm{e}^{-\beta \mathcal{H}_0(\psi_1)} \prod_{i=1}^{n} \langle \psi_{i+1} | \{-V_{b_i}(t_i)\mathrm{d}t_i\} | \psi_i \rangle \,, \tag{3.8}$$

where $\mathcal{H}_0(\psi_1) \equiv \langle \psi_1 | \mathcal{H}_0 | \psi_1 \rangle$.

The status of a variable set $(\{\psi_i\},\{b_i\},\{t_i\})$ is called world-line configuration, because it can be represented as a set of world-lines. Figure 3.1 shows a world-line configuration of an $s = 1/2$ Heisenberg anti-ferromagnet (HAF) model on a four-sites chain.

The X-axis refers the position of spin sites, from 1 to 4, and the Y-axis refers the imaginary time, from 0 to β. In this example, the Hamiltonian consists of local interaction Hamiltonians V_1, V_2 and V_3. The V_b is the Hamiltonian of the HAF interaction between sites b and $b+1$. Then a waving line is drawn at the position corresponding to the status of (b_i, t_i). The vertical line denotes the status of ψ_i between two V_b. In the $s = 1/2$ case, we need two line types, for example, solid and dotted lines which denote up and down spin states of an $s = 1/2$ spin, respectively. In this way, this figure is in one-to-one correspondence with the world-line configuration. Usually, the waving line is called vertex and the position at where a local spin configuration changes is called kink. Since we will only show examples of quantum spin models in the present article, the status of variables ψ_i is called spin configuration.

From (3.7), the function W_n can be regarded as the Boltzmann weight of a world-line configuration. Therefore, using the Matsubara formula, the partition function can be transformed from quantum to classical one.

3.3 Quantum World-line Monte Carlo Method

Using the same technique for the partition function, we can get a similar multiple integral representation of a canonical ensemble average of an observable A.

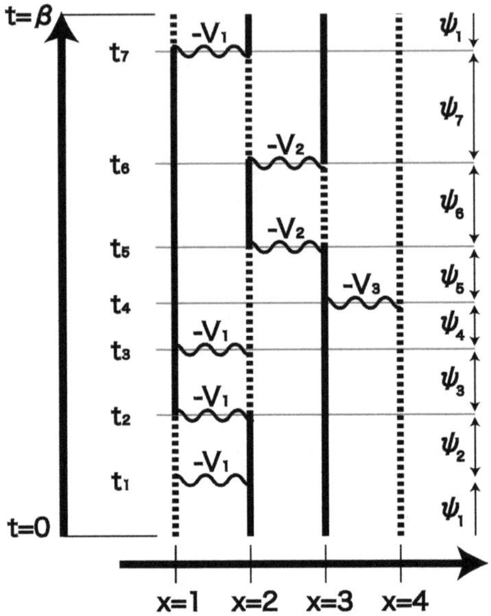

Fig. 3.1. A graphical representation of a world-line configuration for an $s = 1/2$ Heisenberg anti-ferromagnet model on a four-sites chain. *Solid* and *dotted lines* denote spin states $+\frac{1}{2}$ and $-\frac{1}{2}$, respectively. A position of a waiving line corresponds to the status of a variable set (b_i, t_i)

$$\langle A \rangle \equiv \frac{\mathrm{Tr} A\rho(\beta)}{Z} = \sum_{(\{\psi_i\},\{b_i\},\{t_i\})} A(\psi_1) \frac{W_n(\{\psi_i\},\{b_i\},\{t_i\})}{Z} , \tag{3.9}$$

where $A(\psi_1) \equiv \langle \psi_1 | A | \psi_1 \rangle$. Here the symbol $\sum_{(\{\psi_i\},\{b_i\},\{t_i\})}$ denotes the multiple integral for variables $(\{\psi_i\},\{b_i\},\{t_i\})$. In other words, we sum up with respect to all world-line configurations.

Since $\sum_{(\{\psi_i\},\{b_i\},\{t_i\})} W_n = Z$, the function W_n/Z in (3.9) can be regarded as the probability of a world-line configuration. Therefore, if we can sample a world-line configuration with the probability W_n/Z, the canonical ensemble average $\langle A \rangle$ can be calculated as the average of samples of the function $A(\psi_1)$. This is the principle of the QMC method.

Many algorithms have been proposed to make a sample of world-line configurations. However the conventional algorithm before the loop algorithm suffered from some problems. In general, since the probability distribution of world-line configurations is complex, a sampling algorithm of world-line configurations is based on a Markov process. Samples in a Markov process are usually correlated for some time steps which is called correlation time. Unfortunately the correlation time in the conventional sampling algorithm becomes very large near a critical point and at low temperatures, because the size of an updating unit of a world-line configuration is fixed when the correlation length of a quantum system increases. Therefore it is difficult to calculate a

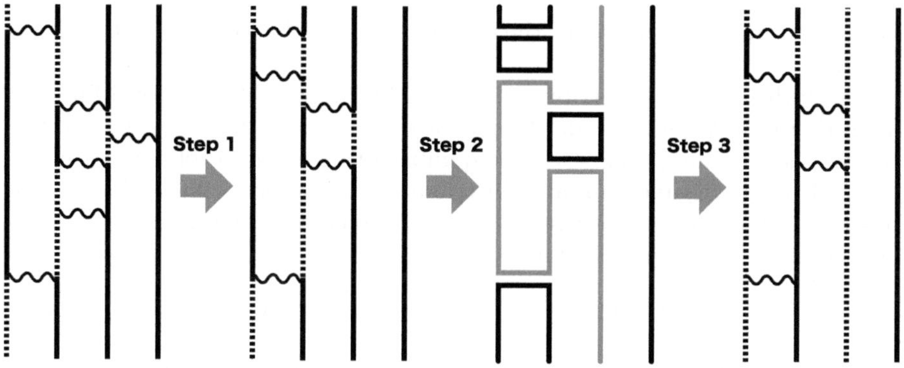

Fig. 3.2. A loop updating for an $s = 1/2$ Heisenberg anti-ferromagnet model on a four-sites chain. *Solid* and *dotted line* denote spin states $+\frac{1}{2}$ and $-\frac{1}{2}$, respectively. The *gray loop* in the third sub-figure is only flipped

canonical ensemble average of a quantum system at interesting points using the conventional algorithm.

3.3.1 Loop Algorithm

The loop algorithm is one of recent developed QMC algorithms. It uses a geometrical object which called loop and updates a world-line configuration with loops globally. For some important quantum models, we can prove that the size of loop in the loop algorithm is equal to the correlation length of the quantum model. This is the reason why the loop algorithm will avoid problems in QMC simulations. In fact, the loop algorithm has some good properties as follows. The first good property is that the correlation time in samples is very short at a critical point and low temperatures. The second one is that various types of world-line configurations can be sampled naturally, which called grand canonical sampling. In practice, it is difficult to change the value of an order parameter without artificial techniques in the conventional algorithm.

The procedure of a loop algorithm consists of three steps. Figure 3.2 illustrates these steps for an $s = 1/2$ HAF model on a four-sites chain. The first step is to update vertexes. In this step, the number of vertexes is changed and those positions are moved, but the spin configuration is fixed. We separately update vertexes and the spin configuration. The second step is to decompose spin variables into loops. In Fig. 3.2 they are decomposed into five loops. The third step is to flip each loop with a probability 1/2. The flipping of a loop is to change the sign of spin variables on the loop. In the quantum $s = 1/2$ spin model, a basis of a quantum state space can be made of the direct product of two states, up and down states, on each spin site. Thus a spin variable takes only two values as $\pm 1/2$ and it can be flipped. Although the gray loop in Fig. 3.2 is only flipped, the spin configuration is drastically changed to the right one. Since there are great differences between the initial and the final world-line configurations, the correlation time of a loop algorithm is very short.

Since the sampling distribution defined by the weight of a world-line configuration is very complex and the partition function is unknown in advance, we use the Markov process whose equilibrium distribution of configurations is equal to the desired sampling distribution. In order to make the Markov process, it is enough to satisfy the detail balance condition for any two configurations A and B in the Markov process:

$$W(\mathrm{A})P(\mathrm{A} \to \mathrm{B}) = W(\mathrm{B})P(\mathrm{B} \to \mathrm{A}), \tag{3.10}$$

where $W(\cdot)$ is the weight function of a desired distribution and $P(\mathrm{A} \to \mathrm{B})$ is the transfer probability from A to B. In other words, the density of random walkers is controlled through their transfer probabilities which balance with the desired sampling distribution. In the loop algorithm, two updatings for vertexes and spin configurations are alternately done and each transfer probability satisfies the detail balance condition.

Firstly, the updating of vertexes is considered. If a vertex V_b is inserted at an imaginary time t between t_{k+1} and t_k ($t_{k+1} \geq t \geq t_k$), the weight of a new world-line configuration W_{n+1} is equal to the present one (W_n in (3.8)) multiplied by a matrix element of V_b with dt: $\langle\psi_{k+1}|\{-V_b(t)\mathrm{d}t\}|\psi_{k+1}\rangle$. The ratio of two weights of pre- and post-inserted a vertex V_b between the imaginary time t and $t + \mathrm{d}t$ is proportional to the infinitesimal dt. Therefore, from (3.10), the accept probability of the inserting of the vertex is proportional to the infinitesimal dt. Such stochastic process is called Poisson process. Scanning a world-line configuration with the Poisson process, we can insert vertexes under the detail balance condition. The thing to do next is to remove a vertex. If a vertex is not on a kink, the ratio of weights between pre- and post-removed configurations is infinite. From (3.10), we can always remove a vertex on a non-kink. On the other hand, if a vertex is on a kink, the ratio of them is exactly zero. Thus it can not be can removed. In summary, the stochastic inserting process of vertexes is a Poisson process whose intensity is equal to a matrix element of a vertex. And the removing a vertex must be done if and only if the vertex is not on a kink.

This updating procedure of vertexes by a Poisson process is applicable to various quantum models, and it is employed in various QMC algorithms. However, on the contrary, it is difficult to find an efficient updating of a spin configuration because of the interaction of spin variables thorough vertexes. In particular, the weight of random generated spin configurations is usually equal to zero, because the matrix element of each vertex often takes a value 0. Thus we need an ingenious updating procedure.

Fortunately, it has been found for some important quantum models. In such cases, the interaction Hamiltonian takes the special form which is called delta operator. The delta operator is defined as the matrix element taking only a value 0 or 1. For example, the HAF interaction Hamiltonian is equal to a special delta operator $\hat{\Delta}(g_H)$ divided by 2 with a special unitary transformation:

$$-\mathcal{H}_{ij}^{\mathrm{HAF}} \equiv -U^{-1}\left(\boldsymbol{S}_i \cdot \boldsymbol{S}_j - \frac{1}{4}\right)U = \frac{1}{2}\hat{\Delta}(g_H)\,. \qquad (3.11)$$

Furthermore, the condition of the matrix element taking a value 1 is as follows:

$$\langle \sigma_i'\sigma_j'|\hat{\Delta}(g_H)|\sigma_i\sigma_j\rangle \equiv \begin{cases} 1 \text{ (if } \sigma_i + \sigma_j = \sigma_i' + \sigma_j' = 0) \\ 0 \text{ (otherwise)} \end{cases}, \qquad (3.12)$$

where $\sigma_i(\sigma_j)$ denotes a spin state on a site $i(j)$ and it takes an eigenvalue of an $s = 1/2$ z-direction spin operator as $\sigma_i \equiv \pm\frac{1}{2}$.

The condition of taking a value 1 for a delta operator can be often represented by a graph. If a delta operator of an interaction Hamiltonian has the graphical representation, we can construct the loop algorithm on the quantum model. For example, the condition of the $\hat{\Delta}(g_H)$ operator is simply represented by a horizontal graph in Fig. 3.3a. The horizontal dotted line in Fig. 3.3a denotes the necessary condition: the sum of two connected spin variables must be equal to zero. From the definition, a matrix element of a delta operator takes only a value 0 or 1. Therefore the value of a vertex weight is not changed, as long as a modification of a spin configuration matches the graphical representation. In the horizontal graph, the modification of the two connected spin variables is only restricted, thus disconnected spin variables are not related to each other: σ_i and σ_i'. Furthermore, the flipping of two connected spin variables together is always allowed, because it satisfies the condition and the weight is not changed. Therefore, we can always flip each set of connected spin variables independently.

In summary, the procedure of updating a spin configuration consists of three steps. In the first step each vertex is replaced with a corresponding graph. The second step is to decompose spin variables into a set which defined by graphs. Since a set of connected spin variables always becomes a loop for the $s = 1/2$ HAF model, this updating algorithm is called loop algorithm. The third step is to flip each loop or set randomly with a probability $\frac{1}{2}$. The loop takes one of two states, because the spin variable takes only a value $\pm\frac{1}{2}$ for the $s = 1/2$ HAF model. Thus we call it *binary loop*.

The loop updating was very successful, because the size of loop is related to the physical correlation length of a quantum model. And by the loop algorithm, many important results were numerically found for various quantum models.

Although we showed only a quantum $s = 1/2$ spin model, the loop updating is also applicable to quantum large spin models. For such cases, Kawashima and Gubernatis proposed a mapping to a quantum $s = 1/2$ spin model which called split-spin technique [5]. The idea is that an $s = m$ spin operator is replaced by the sum of $2m$ $s = 1/2$ spin operators with a permutation operator. Since they are isomorphic, we can simulate a quantum $s = m$ model as a multi quantum $s = 1/2$ model with binary loop updatings. However, since the size of the new quantum space becomes large, we need more memories and complex treatments. In the next section, we will propose a new loop updating which acts on an original quantum space directly.

3.4 Non-binary Loop Algorithm

The symmetry of a model is related to the graphical representation of a vertex. For example, the HAF model has the SU(2) symmetry and the HAF interaction term is represented as the binary horizontal graph. Fortunately, there are such special graphical representations in high-symmetrical cases.

The $s = 1$ bi-linear bi-quadratic (BLBQ) model is interesting theoretically and experimentally [6], because it has some integrable points for a one-dimensional case and it is related to the effective model of Na atoms in an optical lattice [7]. The Hamiltonian is as follows:

$$\mathcal{H}_{ij}^{\mathrm{BLBQ}} \equiv (\cos\theta)\boldsymbol{S}_i \cdot \boldsymbol{S}_j + (\sin\theta)(\boldsymbol{S}_i \cdot \boldsymbol{S}_j)^2 . \tag{3.13}$$

The BLBQ model has SU(3) symmetry at special points, for example, $\theta = -\frac{\pi}{2}$ and $-\frac{3\pi}{4}$. The BLBQ interaction terms at SU(3) points have special graphical representations whose states are non-binary as 0 and ± 1. For example, the BLBQ interaction term at the SU(3) point $\theta = -\frac{\pi}{2}$ becomes a delta operator and the condition taking a value 1 is same to that in an $s = 1/2$ HAF case, except for a spin variable takes three values, ± 1 and 0. As a result, we can use the same horizontal graph in Fig. 3.3a for the BLBQ interaction term. However, the loop, which defined by the SU(3) horizontal graph, takes three states, not binary states.

We can generalize the horizontal graph in the SU(N) case. In the SU(N) case, since the spin variable takes N values ($\sigma_i = \frac{(-N+1)}{2}, \cdots, \frac{(N-1)}{2}$), the number of loop states which defined by the SU(N) horizontal graph is N. Thus we call it *non-binary loop*. Another type of graph can also be generalized in the SU(N). For example, we found the generalized cross graph g_C in Fig. 3.3b which corresponds to the other SU(3) point $\theta = -\frac{3\pi}{4}$:

$$\langle \sigma_i' \sigma_j' | \hat{\Delta}(g_C) | \sigma_i \sigma_j \rangle \equiv \begin{cases} 1 \text{ (if } \sigma_i - \sigma_j' = \sigma_i' - \sigma_j = 0) \\ 0 \text{ (otherwise)} \end{cases} . \tag{3.14}$$

It is easy to check that generalized graphs have the SU(N) symmetry.

The loop defined by the SU(N) graph has N possible states of equal weight, because the exchange of any two states corresponds to one of SU(N) transformations. Therefore we can choose one of N possible states for each loop with equal probability.

In summary, the procedure of the loop updating using the SU(N) graph is similar to that of the conventional loop algorithm with the binary graph. In the first step each vertex is replaced by a corresponding SU(N) graph. And

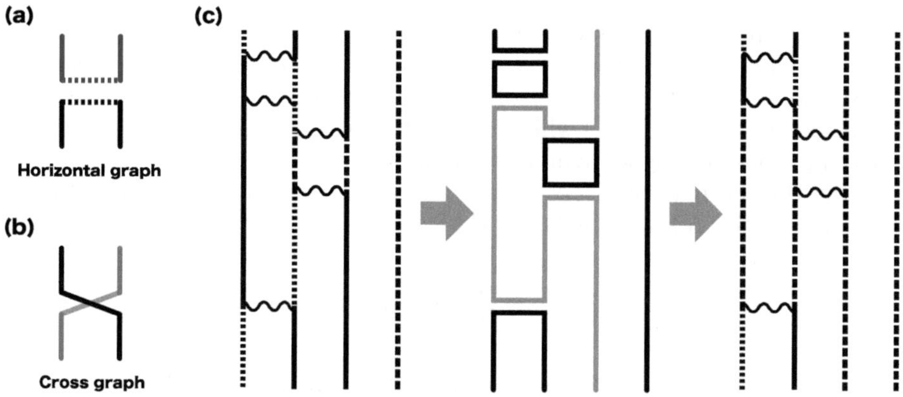

Fig. 3.3. Graphs for various types of vertexes and a non-binary loop updating for an $s = 1$ BLBQ model on a four-sites chain. (**a**) Horizontal graph. (**b**) Cross graph. (**c**) The non-binary loop updating for an $s = 1$ BLBQ model at $\theta = -\frac{\pi}{2}$. *Solid*, *dashed*, and *dotted lines* denote the spin states, +1, 0, and −1, respectively. The gray loop in the second sub-figure is only changed

the second step is to decompose spin variables into a set which defined by the SU(N) graphs. In the third step we choose one of N possible states for each set with equal probability. The only difference between the conventional and new loop algorithms is the number of possible loop states.

The loop algorithm with non-binary loops is called *non-binary loop algorithm* [4]. It is very useful to study high-symmetrical quantum models, because of the simplicity. Without a split-spin technique, we can directly simulate the SU(N) model in the original quantum space. The relation between the conventional and new loop algorithms is similar to that between Swendsen–Wang algorithms for Ising and Potts models.

3.5 Application: Ground States of Two-Dimensional SU(N) Quantum Anti-ferromagnets

Some quantum models already have been simulated with the non-binary loop algorithm. In the present section, we will focus on the numerical study of ground states of two-dimensional SU(N) quantum anti-ferromagnets.

The existence of a short-range resonating valence bond (RVB) spin-liquid is a very interesting problem for low-dimensional quantum systems. A RVB spin-liquid exhibits a finite gap for spin-excitations and it has only short-range order and it does not break any lattice symmetry. Although the RVB spin-liquid may be related to the mechanism of the copper-oxide superconductors, the ground state of the two-dimensional $s = 1/2$ HAF model, which is an effective quantum spin model of copper-oxide superconductor, is not a RVB state, but it has a magnetic order which called Neel order. As the spin s is decreased, the quantum fluctuation decreases the Neel order, but the ground state remains ordered even for $s = 1/2$. Another route to increase quantum fluctuations is in the symmetry of quantum models. We studied the SU(N) quantum anti-ferromagnets which is one of generalizations of the HAF model and which has the higher symmetry than SU(2).

Our Hamiltonian of SU(N) quantum anti-ferromagnets is defined by the generator of SU(N) algebra as the HAF model:

$$\mathcal{H}_{ij}^{\mathrm{SU}(N)} \equiv \frac{1}{N} \sum_{\alpha,\beta=1}^{N} J_{\beta}^{\alpha}(i) J_{\alpha}^{\beta}(j) , \tag{3.15}$$

$$\left[J_{\beta}^{\alpha}(i), J_{\nu}^{\mu}(j)\right] = \delta_{ij} \left(\delta_{\nu}^{\alpha} J_{\beta}^{\mu}(i) - \delta_{\beta}^{\mu} J_{\nu}^{\alpha}(i)\right) , \tag{3.16}$$

where $J_{\beta}^{\alpha}(i)$ is an SU(N) generator on site i. In the SU(2) case, this Hamiltonian is equal to the HAF model. In our anti-ferromagnetic SU(N) model, the representation of SU(N) generator is a N-dimensional matrix which called fundamental representation and the matrix on one sub-lattice is conjugate to that on the other sub-lattice. In this representation, the model can be expressed in terms of SU(2) spins with $s = (N-1)/2$. If the SU(N) Hamiltonian applies to the $|nm\rangle$ state, the result equals the superposition of all $|l\bar{l}\rangle$ states divided by N:

$$\mathcal{H}_{ij}^{\mathrm{SU}(N)} |nm\rangle = -\delta_{n\bar{m}} \left(\frac{1}{N} \sum_{l=-s,\cdots,s} |l\bar{l}\rangle \right) \equiv -\delta_{n\bar{m}} |(ij)\rangle . \quad (\bar{l} \equiv -l) , \tag{3.17}$$

where $|nm\rangle$ denotes the simultaneous eigenstate of z-direction SU(2) spin operators on sites i and j. For convenience, the special superposition state $|(ij)\rangle$ will be called (ij)-bond state in the following.

The (12)-bond state is a ground state of a two-sites case. And it has no magnetic order, because it is the superposition of different magnetic ordered states. In a four-sites case, a bond state $|(12)(34)\rangle$ is an eigenstate of $\mathcal{H}_{12}^{\mathrm{SU}(N)}$ and $\mathcal{H}_{34}^{\mathrm{SU}(N)}$ whose eigenvalue is -1. However, if the interaction Hamiltonian does not match the covering of dimers over sites in a bond state, the state converges to a zero-eigenstate when N becomes infinity:

$$\mathcal{H}_{23}^{\mathrm{SU}(N)} |(12)(34)\rangle = -\frac{1}{N} |(41)(23)\rangle \to 0 \quad (N \to \infty) . \tag{3.18}$$

Therefore, in the large N limit, the ground state maximizes the number of nearest-neighbor bonds, because the eigenvalue of a whole Hamiltonian is minus of the number of nearest-neighbor bonds. In short, in the large N limit, due to strong quantum fluctuations, the ground state has no magnetic order. However, in a one-dimensional chain, the number of maximum bond state is only two [9]. Since they break the translational symmetry, they are not a RVB state, which called dimer state. Read and Sachdev studied the ground state of SU(N) models on a two-dimensional square lattice theoretically [8]. Their conclusion is also that the ground state on a two-dimensional square lattice is a dimer state in the large N limits.

In the $N = 2$ case, the ground state is a magnetic ordered Neel state. On the other hand, in the large N limit, it has no magnetic order and it becomes a dimer state by Read and Sachdev's prediction. However the intermediate N cases remain problems. The numerical study by Santoro et al. suggested a RVB spin-liquid ground state in the $N = 4$ case [10]. The model of Santoro et al. is equal to our SU(4) model and it also appears as a special point in coupled spin-orbital models. Santoro et al. used the Green function Monte Carlo method in their numerical simulations. Due to the computational times by their numerical method, the size of their simulated lattices was limited to only $L = 12$.

Since the SU(N) interaction Hamiltonian is equal to the delta operator $\hat{\Delta}(g_H)$ represented by the non-binary horizontal graph, this model can be simulated for various N and large square lattices by a non-binary loop algorithm. Indeed, we explored the system size L up to 128 and N up to 8 [11]. Our simulations have been performed at low enough temperatures to be effectively in the ground state. The lowest temperature for $N = 4$ and $L = 128$, for example, is $\frac{1}{128}$.

In order to search the order of a ground state, we have measured two quantities which are related to magnetic and dimer orders, respectively. The one is a static structure factor $S(\pi, \pi)$ of staggered magnetization m_s and the other is a dimer static structure factor $S^{\mathrm{D}}(\pi, 0)$:

$$\frac{S(\pi, \pi)}{L^2} \equiv \langle (m_s)^2 \,, \rangle \quad \left(m_s \equiv \frac{1}{L^d} \sum_r (-1)^r S_r^z \right) \tag{3.19}$$

and

$$\frac{S^{\mathrm{D}}(\pi, 0)}{L^2} \equiv \langle (d_{(\pi,0)})^2 \rangle \,, \quad \left(d_{(\pi,0)} \equiv \frac{1}{L^d} \sum_r (-1)^{r_x} S_r^z S_{r+x}^z \right) \,. \tag{3.20}$$

If an order exists, the related structure factor divided by L^2 converges to a finite value in the large-L limits, but if the order is absent, it decreases to zero as $1/L^2$.

Figure 3.4 shows a clear evidence for a Neel order in $N \leq 4$, because the spin structure factor divided by L^2 converges to a finite value for $N \leq 4$. Therefore the ground state is not a RVB spin-liquid even in the $N = 4$ case. On the contrary, in the $N \geq 5$ cases, the spin structure factor divided by L^2 decreases as $1/L^2$. Therefore there is no Neel order in $N \geq 5$. The disagreement of our conclusion for the $N = 4$ case with that of Santoro et al. is not in the raw numerical data. Their data are only in the $L \leq 12$ region. Therefore, the disagreement is solely due to the small system sizes studied in [10].

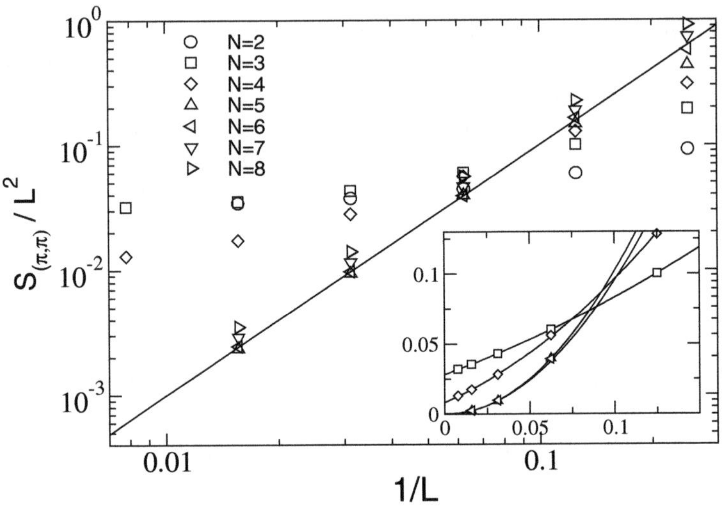

Fig. 3.4. Static structure factors $S_{(\pi,\pi)}$ for $2 \leq N \leq 8$. The *straight line* representing the power law L^{-2} is drawn for comparison. Estimated statistical errors are not shown, because they are equal to or smaller than the symbol size. The *inset* presents the data for $N = 3, 4, 5, 6$ in the linear scale, together with the best fitting curves obtained by the method of least squares. (Adopted from Harada and Kawashima [11])

On the other hand, Fig. 3.5 shows a clear evidence for a dimer order in $N \geq 5$. While the dimer structure factor divided by L^2 decreases to zero in $N = 4$, it converges to a finite value in $N \geq 5$. Therefore the dimer order exists in $N \geq 5$ and the ground state is not a RVB spin-liquid state. It is consistent with the theoretical prediction by Read and Sachdev in the large N limits.

In summary, from QMC simulations for large square lattices by the loop algorithm with non-binary loops, we confirmed a direct transition from the Neel ground state ($N \leq 4$) to the dimer ground state ($N \geq 5$) for two-dimensional SU(N) quantum anti-ferromagnets. These comprehensive numerical studies can be effectively done by the loop algorithm with non-binary loops.

3.6 Conclusion

We have introduced the QMC method with non-binary updatings for quantum models with high symmetry. For example, using graphical representations of the interaction Hamiltonian, we can easily simulate a high-symmetrical quantum model by a loop algorithm with non-binary loops. We have only focused to the loop algorithm, but the idea of a non-binary updating for quantum models with high symmetry can be combined with the other algorithms as the worm

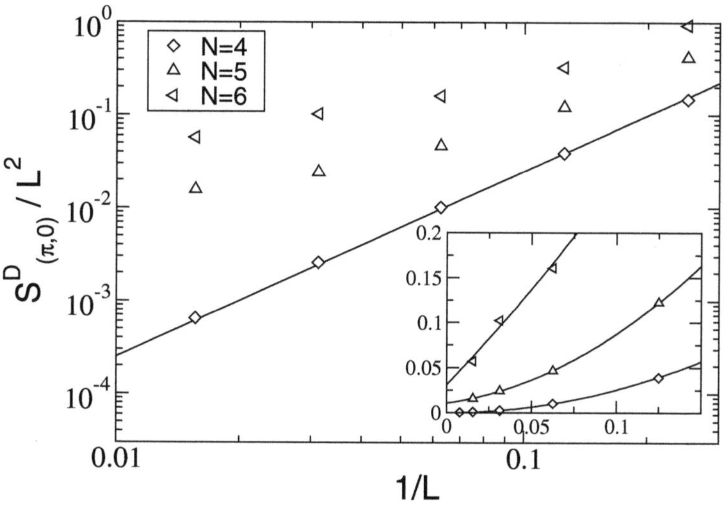

Fig. 3.5. The $\boldsymbol{k} = (\pi, 0)$ dimer structure factors $S^{\mathrm{D}}_{(\pi,0)}$ for $N = 4, 5, 6$ in logarithmic scale. The *inset* is the linear-scale plot. The *solid lines* in the inset are the best fitting curves of least squares based on the $L \gtrsim 8$ data. (Adopted from Harada and Kawashima [11])

or directed loop. We have shown one example for using a non-binary loop algorithm. It is the problem of the ground state of two-dimensional SU(N) quantum anti-ferromagnets. Our conclusion is that our SU(N) models have no RVB spin-liquid ground state.

Acknowledgments

The author is grateful to N. Kawashima and Y. Okabe for stimulating conversations and useful comments.

References

1. M. Suzuki: Prog. Theor. Phys. **56**, 1454 (1976)
2. H.G. Evertz, G. Lana, M. Marcu: Phys. Rev. Lett. **70**, 875 (1993)
3. H.G. Evertz: Advances in Physics **52**, 1 (2003)
4. N. Kawashima, K. Harada: J. Phys. Soc. Jpn. **73**, 1379 (2004)
5. N. Kawashima, J.E. Gubernatis: Phys. Rev. Lett. **73**, 1295 (1994)
6. For physical results on this model, see, for example, A. Schmitt, K-H. Mütter, M. Karbach, Y. Yu, G. Müller: Phys. Rev. B **58**, 5498 (1998), and references cited therein; also see K. Nomura, S. Takada: J. Phys. Soc. Jpn. **60**, 389 (1991)
7. M. Greiner, O. Mandel, T. Esslinger, T.W. Hänsch, I. Bloch: Nature **415**, 39 (2002)
8. N. Read, S. Sachdev: Phys. Rev. B **42**, 4568 (1990)

9. I. Affleck: Phys. Rev. Lett. **54**, 966 (1985)
10. G. Santoro, S. Sorella, L. Guidoni, A. Parola, E. Tosatti: Phys. Rev. Lett. **83**, 3065 (1999)
11. K. Harada, N. Kawashima: Phys. Rev. Lett. **90**, 117203 (2003)

Event-by-Event Simulation of Quantum Phenomena

H. De Raedt[1], K. De Raedt[2], and K. Michielsen[1]

[1] Applied Physics – Computational Physics, Materials Science Centre, University of Groningen, Nijenborgh 4, NL-9747 AG,Groningen, The Netherlands
[2] Department of Computer Science, University of Groningen, Blauwborgje 3, NL-9747 AC Groningen, The Netherlands

Abstract. In various basic experiments in quantum physics, observations are recorded event-by-event. The final outcome of such experiments can be computed according to the rules of quantum theory but quantum theory does not describe single events. In this paper, we describe a stimulation approach that does not rely on concepts of quantum theory but nevertheless generates events with frequencies that agree with quantum theory. In particular, we demonstrate that locally connected networks of processing units that have primitive learning capabilities can be used to perform a deterministic, event-based simulation of single-photon Mach-Zehnder interferometer experiments.

4.1 Introduction

Recent advances in nanotechnology, make it possible to control individual ions, atoms, photons and the like. These technological developments facilitate the study of single quantum systems at the level of individual events [1, 2]. Such experiments address the most fundamental aspects of quantum theory. Quantum theory gives a recipe to compute the frequencies for observing events but it does not describe individual events, such as the arrival of a single electron at a particular position on the detection screen [2–5]. Reconciling the mathematical formalism, not describing single events, with the experimental fact that each observation yields a definite outcome is often referred to as the quantum measurement paradox [3, 4].

Computer simulation is a powerful methodology, complementary to theory and experiment, to model physical phenomena [6]. In this approach, we start from the basic equations of physics and use numerical algorithms to solve these equations. But what if the basic equation to describe the individual events is unknown?

From a computational viewpoint, quantum theory provides a set of rules to compute probability distributions [3, 7], that is to compute the final, collective

Springer Proceedings in Physics, Volume 105
Computer Simulation Studies in Condensed-Matter Physics XVIII
Eds.: D.P. Landau, S.P. Lewis and H.-B. Schüttler

outcome of an experiment in which single events are recorded. However, quantum theory does not provide algorithms to perform an event-based simulation of such experiments. Methods based on the solution of the (time-dependent) Schrödinger equation are inappropriate for this purpose. Hence, a completely new computer simulation method is required to simulate the event-based observations in quantum physics experiments.

Elsewhere, we have already demonstrated that locally-connected networks of processing units with a primitive learning capability can generate events at a rate that agrees with the quantum mechanical probability distribution [8–11]. In this paper we discuss the basic elements of our approach and, as an illustration, we present deterministic event-based simulation results of single-photon Mach–Zehnder interferometer experiments.

4.2 Deterministic Learning Machines (DLMs)

In quantum physics, an event corresponds to the detection of a photon, electron or the like. In our deterministic, event-based simulation approach an event is the arrival of a message at the input port of a processing unit. This processing unit typically contains three components: an input DLM, a transformation unit and an output DLM. A DLM is a classical (but non-Hamiltonian), deterministic, local and causal dynamical system with a primitive learning capability. A DLM consists of input and output ports, an internal unit vector, a rule that specifies how this vector changes when an input event is received, and a rule by which the DLM determines the type of output event it generates as a response to the input event.

For simplicity, we assume that the input DLM can accept $L = 2$ types of events, but only one at a time. We label the events by 0 and 1. Event 0 carries a message represented by a unit vector $\boldsymbol{y}_0 = (y_0, y_1)$ containing two real numbers. Similarly, event 1 carries a message represented by $\boldsymbol{y}_1 = (y_2, y_3)$. In this particular case the internal vectors of the input and output DLM have length four and can be represented by $\boldsymbol{x} = (x_0, x_1, x_2, x_3)$, where x_i, $i = 0, \ldots, 3$ are real numbers. In general the length of the internal vector is 2×2^L. The initial value of the internal vector is irrelevant and can be chosen at random.

The learning algorithm of the input DLM, DLMi, is defined as follows:

- DLMi constructs a vector $\hat{\boldsymbol{x}}$ of length four using information from the incoming event and from its own internal vector $\boldsymbol{x}$. If the DLM receives a 0 event then $\hat{\boldsymbol{x}} = (y_0, y_1, x_2, x_3)$. If it receives a 1 event then $\hat{\boldsymbol{x}} = (x_0, x_1, y_2, y_3)$.
- Based on its own internal vector DLMi computes eight candidate internal vectors containing the elements

$$w_{j,j} = \pm\sqrt{1 + \alpha^2(x_j^2 - 1)}\,, \quad w_{j,i} = \alpha x_i\,, \quad \text{if} \quad i \neq j\,, \tag{4.1}$$

where $i, j = 0, \ldots, 3$ and i is the first running index. The parameter $0 < \alpha < 1$ controls the learning process.

- DLMi chooses from the eight candidate internal vectors $\boldsymbol{w}$ the one that minimizes the cost function $C = -\boldsymbol{w}.\hat{\boldsymbol{x}}$ and replaces its internal vector $\boldsymbol{x}$ by this vector.
- DLMi puts the four elements of its internal vector on its four output ports and waits for the next event to be processed.
- The four output ports of DLMi are connected to the four input ports of the transformation unit. The transformation unit applies an orthogonal transformation to the vector $\boldsymbol{x} = (x_0, x_1, x_2, x_3)$ it receives from DLMi. The precise form of the transformation depends on the particular function that the processor has to perform. An example is given in Sect. 4.4. The result of the transformation is the vector $\boldsymbol{x}' = (x'_0, x'_1, x'_2, x'_3)$.

The learning algorithm of the output DLM, DLMo, is defined as follows:

- DLMo takes the vector $\boldsymbol{x}' = (x'_0, x'_1, x'_2, x'_3)$ as input.
- Just as DLMi, DLMo computes, based on its internal vector eight, candidate internal vectors according to (4.1).
- DLMo chooses from the eight candidate internal vectors $\boldsymbol{w}$ the one that minimizes the cost function $C = -\boldsymbol{w}.\boldsymbol{x}'$ and replaces its internal vector $\boldsymbol{x}$ by this vector.
- DLMo generates an output event of type 0 (1) carrying as a message the first (last) two elements of its new internal vector, if an update rule with $j = 0, 1$ ($j = 2, 3$) was chosen.
- DLMo waits for the next event to be processed.

4.3 A DLM is an Efficient Encoder

We consider as a special case of the DLM described in Sect. 4.2 a DLM which accepts one input event carrying the message $\boldsymbol{y} = (\cos\phi, \sin\phi)$. In this case, the internal vector $\boldsymbol{x}$ of the DLM has length two and there are four candidate update internal vectors, given by (4.1) with $i, j = 0, 1$. The DLM chooses from the four candidate internal vectors $\boldsymbol{w}$ the one that minimizes the cost function $C = -\boldsymbol{w}.\boldsymbol{y}$ and replaces its internal vector $\boldsymbol{x}$ by this vector. The DLM sends out a 0 (1) event carrying as a message the first (second) element of its new internal vector if an update rule with $j = 0$ ($j = 1$) was chosen. If we count the number of 1 events sent out by the DLM, which we denote by K, and divide by the total number of processed events N, we find $K/N \approx \cos^2\phi$. The DLM generates a fully deterministic sequence of zeroes and ones, that is, the most compact sequence for each K and N, with minimum variance on $K/N \approx \cos^2\phi$. The performance is optimal: the number of distinguishable messages is equal to $N + 1$.

Apparently, the most efficient deterministic encoder (encoding an angle ϕ as a sequence of zeroes and ones) that we can build generates data according

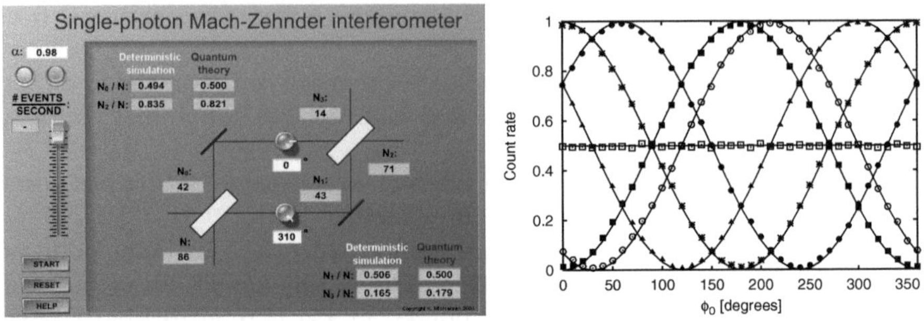

Fig. 4.1. *Left*: Screen dump of an interactive program that performs event-based simulations of a single-photon Mach–Zehnder interferometer [12]. The number of events N_i in channel $i = 0, \ldots, 3$ corresponds to the probability for finding a photon on the corresponding arm of the interferometer. *Right*: Simulation results for the DLM-network shown on the left. Input channel 0 receives $(\cos\psi_0, \sin\psi_0)$ with probability one. A uniform random number in the range $[0, 360]$ is used to choose the angle ψ_0. Input channel 1 receives no events. Each data point represents 10000 events ($N_0 + N_1 = N_2 + N_3 = 10000$). Initially the rotation angle $\phi_0 = 0$ and after each set of 10000 events, ϕ_0 is increased by 10°. Markers give the simulation results for the normalized intensities as a function of ϕ_0. *Open squares*: $N_0/(N_0+N_1)$; *solid squares*: $N_2/(N_2+N_3)$ for $\phi_1 = 0$; *open circles*: $N_2/(N_2+N_3)$ for $\phi_1 = 30°$; *bullets*: $N_2/(N_2+N_3)$ for $\phi_1 = 240°$; *asterisks*: $N_3/(N_2+N_3)$ for $\phi_1 = 0$; *solid triangles*: $N_3/(N_2+N_3)$ for $\phi_1 = 300°$. *Lines* represent the results of quantum theory

to Malus's law: $I = I_0 \cos^2\phi$, where ϕ is the angle between the polarization direction of the light and the transmission axis of the polarizer. Note that we designed the encoder by making use of geometric rules and not by using laws of physics. The $\cos^2\phi$ law is the result of finding the optimal encoder.

4.4 Single-Photon Mach–Zehnder Interferometer

We consider a DLM network (see Fig. 4.1) that generates the same interference patterns as those observed in single-photon Mach–Zehnder interferometers [1]. There is a one-to-one correspondence between the processing units in the network and the physical parts of the experimental setup. For the beam splitters we use a network consisting of an input and output DLM as described in Sect. 4.2. To implement the operation of a beam splitter, we use in the transformation unit the transformation $(x_0, x_1, x_2, x_3) \to 1/\sqrt{2}(x_0 - x_3, x_2 + x_1, x_2 - x_1, x_0 + x_3)$. The phase shift is taken care of by two passive devices that perform plane rotations by ϕ_0 and ϕ_1, respectively. According to quantum theory, in the case that input channel 1 receives no input events, the amplitudes (b_0, b_1) of the photons in the output channels N_2 and N_3 of the Mach-Zehnder interferometer are given by

$$b_0 = e^{i\gamma}\sin((\phi_0 - \phi_1)/2)\,, \quad b_1 = e^{i\gamma}\cos((\phi_0 - \phi_1)/2)\,, \qquad (4.2)$$

where γ is an irrelevant phase. From (4.2), it follows that the probabilities $|b_0|^2$ and $|b_1|^2$ depend on $\phi = \phi_0 - \phi_1$ only. In Fig. 4.1 we present a representative selection of simulation results for the Mach-Zehnder interferometer built from DLMs. It is clear that our event-by-event simulation approach generates events according to the wave mechanical distribution (4.2).

4.5 Discussion

We have described the basic elements of a procedure to construct algorithms that can be used to simulate quantum processes without solving the Schrödinger equation. We have shown that single-particle quantum interference can be simulated on an event-by-event basis using local and causal processes, without using concepts such as wave fields or particle-wave duality. Elsewhere, we show that the same procedure can be used to perform a deterministic, event-based simulation of universal quantum computation [11, 13]. These results suggest that we may have discovered algorithms that do not rely on concepts of quantum theory, but are able to simulate quantum phenomena using classical, causal, local, and event-based processes.

References

1. P. Grangier, R. Roger, A. Aspect: Europhys. Lett. **1**, 173 (1986)
2. A. Tonomura: *The Quantum World Unveiled by Electron Waves.* (World Scientific, Singapore 1998)
3. D. Home: *Conceptual Foundations of Quantum Physics.* (Plenum Press, New York 1997)
4. R.P. Feynman, R.B. Leighton, M. Sands: *The Feynman lectures on Physics.* Vol. 3 (Addison-Wesley, Reading MA 1996)
5. L.E. Ballentine: *Quantum Mechanics: A Modern Development.* (World Scientific, Singapore 2003)
6. D.P. Landau, K. Binder: *A Guide to Monte Carlo Simulation in Statistical Physics.* (Cambridge University Press, Cambridge 2000)
7. N.G. Van Kampen: Physica A **153**, 97 (1988)
8. K. De Raedt, H. De Raedt, K. Michielsen: arXiv: quant-ph/0409213
9. H. De Raedt, K. De Raedt, K. Michielsen: Europhys. Lett. **69**, 861 (2005)
10. H. De Raedt, K. De Raedt, K. Michielsen: J. Phys. Soc. Jpn. (in press)
11. K. Michielsen, K. De Raedt, H. De Raedt: J. Comp. Theor. Nanoscience (in press)
12. For additional information visit http://www.compphys.net/dlm
13. K. Michielsen, K. De Raedt, H. De Raedt: *Deterministic event-based simulation of universal quantum computation.* In: Computer Simulation Studies in Condensed-Matter Physics XVIII, D.P. Landau et al. (Eds.), Springer Proceedings in Physics, Vol. 105 (Springer, Berlin Heidelberg New York 2006)

5

Deterministic Event-based Simulation of Universal Quantum Computation

K. Michielsen[1], H. De Raedt[1], and K. De Raedt[2]

[1] Applied Physics – Computational Physics, Materials Science Centre, University of Groningen, Nijenborgh 4, NL-9747 AG, Groningen, The Netherlands
[2] Department of Computer Science, University of Groningen, Blauwborgje 3, NL-9747 AC, Groningen, The Netherlands

Abstract. We demonstrate that locally connected networks of classical processing units that have primitive learning capabilities can be used to perform a deterministic, event-based simulation of universal quantum computation. The new simulation method is applied to implement Shor's factoring algorithm.

5.1 Introduction

The basic ideas of quantum computation were formulated more than twenty years ago [1,2]. Ten years ago DiVincenzo has proven that the CNOT gate and single-qubit operations constitute a set of gates that can be used to construct a universal quantum computer [3]. This statement is equivalent to the one that a digital classical computer can be constructed by means of NAND gates only. Conventional computers can simulate the physical behavior of quantum computer hardware by solving the (time-dependent) Schrödinger equation [4, 5]. However, the idea of building a quantum computer has lead to many advances in nanotechnology, making it possible to control individual ions, atoms, photons and the like and these single events cannot be described by quantum theory. In this paper we demonstrate how a quantum computer can be built from locally-connected networks of processing units with a primitive learning capability that deterministically generate events with frequencies that agree with the corresponding quantum mechanical probabilities.

5.2 Stochastic Learning Machines (SLMs)

In [6] we explained in detail how a DLM works. Here we describe its stochastic variant. The term stochastic refers to the method that is used to select the output channel that will carry the outgoing message. Only changes are made to the learning algorithm of the output DLM, DLMo, described in [6]. It can

Springer Proceedings in Physics, Volume 105
Computer Simulation Studies in Condensed-Matter Physics XVIII
Eds.: D.P. Landau, S.P. Lewis and H.-B. Schüttler

be proven that in the stationary regime $x_0^2+x_1^2$ and $x_2^2+x_3^2$, where x_0, x_1, x_2 and x_3 are the four elements of the internal vector $\boldsymbol{x}$ of a DLM, correspond to the probabilities of quantum theory. Instead of sending out messages in a deterministic way as described in [6], we choose a random number $0<r<1$ and send out a zero event if $x_1^2+x_2^2 \leq r$ and a one event otherwise. Although the learning process of this processor is still deterministic, in the stationary regime the output events are randomly distributed over the two possibilities. Replacing DLMs by SLMs in a DLM-network changes the order in which messages are being processed but leaves the content of the messages intact.

5.3 Universal Quantum Computation: Concepts

Qubit. The state $|\Phi\rangle$ of a qubit can be written as $|\Phi\rangle = a_0|0\rangle + a_1|1\rangle$, where a_0 and a_1 are complex numbers so that $|a_0|^2+|a_1|^2=1$. In general, the state $|\Phi\rangle$ of the qubit can also be written as $|\Phi\rangle = \sqrt{p_0}e^{i\psi_0}|0\rangle + \sqrt{p_1}e^{i\psi_1}|1\rangle$, where p_0 and $p_1 = 1-p_0$ denote the probability to find the qubit in state $|0\rangle$ and $|1\rangle$, respectively, and ψ_0 and ψ_1 denote phase factors.

In principle, any physical object that carries a $S=1/2$ degree-of-freedom can serve as a physical realization of a qubit [7]. The state $|\phi\rangle$ of a qubit can therefore also be written as a linear combination of the spin-up and spin-down states [8] $|\Phi\rangle = a_0|\uparrow\rangle + a_1|\downarrow\rangle$, where $|\uparrow\rangle = |0\rangle = (10)^T$ and $|\downarrow\rangle = |1\rangle = (01)^T$ [7]. The three components of the spin-1/2 operator $\boldsymbol{S} = (S^x, S^y, S^z)$ are defined (in units such that $\hbar = 1$) by [8]

$$S^x = \frac{1}{2}\begin{pmatrix} 0 & 1 \\ 1 & 0 \end{pmatrix}, \quad S^y = \frac{1}{2}\begin{pmatrix} 0 & -i \\ i & 0 \end{pmatrix}, \quad S^z = \frac{1}{2}\begin{pmatrix} 1 & 0 \\ 0 & -1 \end{pmatrix}, \tag{5.1}$$

and have been chosen such that $|\uparrow\rangle$ and $|\downarrow\rangle$ are eigenstates of S^z with eigenvalues $+1/2$ and $-1/2$, respectively. The expectation values of the three components of the qubits are defined as $\langle Q^\alpha\rangle = 1/2 - \langle S^\alpha\rangle$, $\alpha = x,y,z$, where $\langle A\rangle = \langle\Phi|A|\Phi\rangle/\langle\Phi|\Phi\rangle$. A qubit is in the state $|0\rangle$ or $|1\rangle$ if $\langle Q^z\rangle = 0$ or $\langle Q^z\rangle = 1$, respectively.

Single Qubit Operations. The most general operation on a single qubit can be expressed as a rotation of the operator $\boldsymbol{S}$ about a vector $\boldsymbol{v}$

$$e^{i\boldsymbol{v}\cdot\boldsymbol{S}} = \mathbf{1}\cos\frac{v}{2} + \frac{2i\boldsymbol{v}\cdot\boldsymbol{S}}{v}\sin\frac{v}{2}, \tag{5.2}$$

where $\mathbf{1}$ denotes the unit matrix and v is the length of the vector $\boldsymbol{v}$. A special case of (5.2) is the Hadamard operation defined by [7]

$$H \equiv e^{-i\pi/2}e^{i\pi(S^x+S^z)/2} = \frac{1}{\sqrt{2}}\begin{pmatrix} 1 & 1 \\ 1 & -1 \end{pmatrix}. \tag{5.3}$$

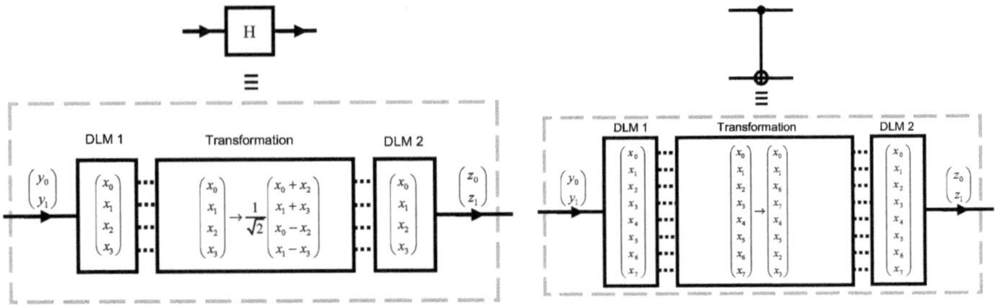

Fig. 5.1. Diagram of DLM networks that perform a deterministic event-based simulation of a Hadamard gate (*left*) and a CNOT gate (*right*). The arrows on the solid lines represent the input and output events. Dashed lines indicate the flow of data within the DLM-based processor

CNOT Operation. By definition the CNOT gate flips the target qubit if the control qubit is in the state $|1\rangle$. If we take the first qubit, that is the least significant bit in the binary notation of an integer, as the control bit the operation of the CNOT gate on a two qubit state $|\phi\rangle = a_0|00\rangle + a_1|01\rangle + a_2|10\rangle + a_3|11\rangle$ results in $|\phi'\rangle = a_0|00\rangle + a_3|01\rangle + a_2|10\rangle + a_1|11\rangle$. In other words, the probability amplitudes of the states $|01\rangle$ and $|11\rangle$ interchange.

5.4 Hadamard and CNOT Gate: DLM-based Processors

The diagram of a network that performs a deterministic event-based simulation of a Hadamard gate is shown in Fig. 5.1left. The DLM-based Hadamard gate consists of three units. Unit one, called DLM1, "learns" about the occurrence of 0 and 1 events, as described in [6]. The 0 event, corresponding to a qubit in state $|0\rangle$, carries a message $\boldsymbol{y_0} = (y_0, y_1) = (\cos\psi_0, \sin\psi_0)$ and the 1 event, corresponding to a qubit in state $|1\rangle$, carries a message $\boldsymbol{y_1} = (y_2, y_3) = (\cos\psi_1, \sin\psi_1)$. Unit two transforms the output of DLM1 by performing an Hadamard operation and feeds this data in the third unit, called DLM2. DLM2 "learns" about the transformed data and responds to the input event by sending out either a 0 or 1 event, as described in [6].

In the quantum system a Hadamard gate transforms the qubit state $|\Phi\rangle$ into $|\Phi'\rangle = [(a_0 + a_1)|0\rangle + (a_0 - a_1)|1\rangle]/\sqrt{2}$. In the corresponding classical system the ys are thus transformed as $(y_0, y_1) \to (y_0 + y_2, y_1 + y_3)/\sqrt{2}$ and $(y_2, y_3) \to (y_0 - y_2, y_1 - y_3)/\sqrt{2}$. This is the transformation performed by the second unit in the DLM-based Hadamard processor.

The schematic diagram of the DLM-network that performs the CNOT operation on an event-by-event (particle-by-particle) basis is shown in Fig. 5.1 right. Conceptually the structure of the network is the same as in the case of a single qubit operation. We now have four instead of two different types of input events, corresponding to the quantum states $|00\rangle$, $|01\rangle$, $|10\rangle$ and $|11\rangle$. Each event carries a message consisting of two real numbers $\boldsymbol{y} = (\cos\phi_i, \cos\phi_i)$ for $i = 0, \ldots, 3$, corresponding to the phase of the quantum mechanical probability amplitudes $a_0/|a_0|$, $a_1/|a_1|$, $a_2/|a_2$ and $a_3/|a_3|$, respectively. The internal unit vector of the input and output DLMs have length eight and there are sixteen candidate update rules. The transformation in the transformation unit is simple: all it has to do is swap the two pairs of elements (x_2, x_3) and (x_6, x_7). This corresponds to the interchange of the probability amplitudes of the states $|01\rangle$ and $|11\rangle$ in the quantum system.

Summarizing, by making use of the DLM-networks for the Hadamard and CNOT gate we can build a deterministic event-based universal quantum computer.

5.5 Factoring $N = 15$ Using Shor's Algorithm

As an example of a quantum algorithm running on the deterministic event-based quantum computer we consider the problem of factoring $N = 15$ on a seven-qubit quantum computer using Shor's algorithm [7,9]. Shor's algorithm is based on the fact that the factors p and q of an integer $N = pq$ can be deduced from the period M of the function $f(j) = a^j \mathrm{mod} N$ for $j = 0, \ldots, 2^n - 1$ where $N \leq 2^n$. Here $a < N$ is a random number that has no common factors with N. Once M has been determined at least one factor of N can be found by computing the greatest common divisor of N and $a^{M/2} \pm 1$. The quantum network for $N = 15$, $a = 11$ can be found in [10]. The quantum network contains Hadamard and CNOT gates and a network to perform the Fourier transform, containing Hadamard gates and controlled phase shifts. In this particular case the period M of the function $f(j)$ can be determined from the expectation values of the first three qubits, that are the qubits involved in the Fourier transform. According to quantum theory we expect to find $Q_1 = \langle Q_1^z \rangle = 0$, $Q_2 = \langle Q_2^z \rangle = 0.5$ and $Q_3 = \langle Q_3^z \rangle = 0.5$.

We transform the quantum circuit in a DLM or SLM circuit by replacing the quantum gates by DLM or SLM-based gates. The DLM or SLM network to perform a controlled phase shift is constructed by mimicking the procedure for constructing the CNOT gate. We count the number of 1 events in the three output channels of the Fourier transform and divide these numbers by the total number of events analyzed to obtain numerical estimates for the qubits Q_1, Q_2 and Q_3. In Fig. 5.2 we present simulation results for the DLM (left) and SLM (right) implementation of the circuit. After processing a few events the results of quantum theory are reproduced with high accuracy.

5.6 Discussion

We have demonstrated that networks of locally connected processing units with primitive learning capabilities are capable of simulating universal quan-

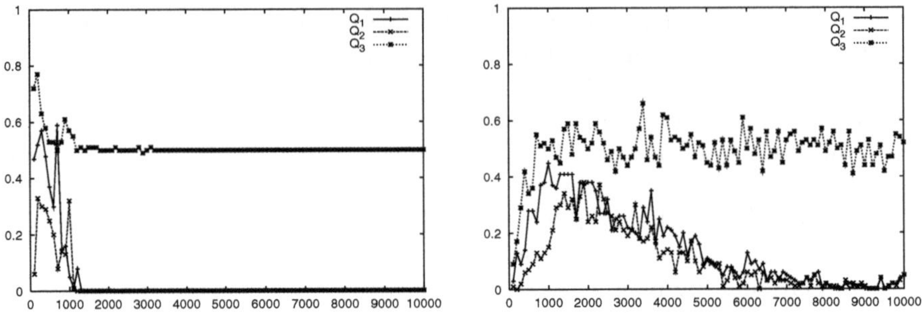

Fig. 5.2. Event-based simulation of Shor's quantum algorithm for factoring $N = 15$, using the value $a = 11$. Each data point represents the average of 100 output events. The parameter that controls the learning process is $\alpha = 0.999$. *Left*: Results for DLM-network. *Right*: Results for SLM-network

tum computers, not through solving the Schrödinger equation, but by generating event by event. Since it is known that the time evolution of the wave function of a quantum system can be simulated on a quantum computer [7], it should thus be possible to simulate real-time quantum dynamics through a deterministic event-based simulation. In conclusion, the work presented in this paper suggests that there exist deterministic, particle-like processes that generate the probability distributions of quantum theory.

References

1. P. Benioff: J. Stat. Phys. **22**, 563 (1980).
2. R.P. Feynman: Int. J. Theor. Phys. **21**, 467 (1982)
3. D.P. DiVincenzo: Phys. Rev. A **51**, 1015 (1995)
4. H. De Raedt, A.H. Hams, K. Michielsen, K. De Raedt: Comp. Phys. Comm. **132**, 1 (2000)
5. H. De Raedt, K. Michielsen, in *Handbook of Computational and Theoretical Nanotechnology.* Edited by M. Rieth and W. Schommers (American Scientific Publisher, Los Angeles 2005) in press
6. H. De Raedt, K. De Raedt, K. Michielsen: *Event-by-event simulation of quantum phenomena.* In: Computer Simulation Studies in Condensed-Matter Physics XVIII, D.P. Landau et al. (Eds.) Springer Proceedings in Physics, Vol. 105 (Springer, Berlin Heidelberg New York 2006)
7. M.A. Nielsen, I.L. Chuang: *Quantum computation and quantum information.* (Cambridge University Press, Cambridge, New York 2000)
8. G. Baym: *Lectures on Quantum Mechanics.* (W.A. Benjamin, Reading MA, 1974)
9. P.W. Shor: SIAM Review **41**, 303 (1999)
10. U.M.K. Vandersypen, M. Steffen, G. Breyta, C.S. Yannoni, M.H. Sherwood, I.L. Chuang: Nature **414**, 883 (2001)

Part II

Materials and Nanostructures

6

Unstable Crack Motion is Predictable

F.F. Abraham

The University of Georgia, Athens GA 30602, USA
Graham-Purdue Visiting Professor of Physics

Abstract. Yoffe's linear theory of dynamic brittle fracture suggests that crack motion will be unstable beyond $\approx 70\%$ of the Rayleigh speed, a prediction that is not supported by experiment. We show by atomistic simulations that hyperelasticity, the elasticity of large strains, plays a governing role in the instability dynamics of brittle fracture. A simple, yet remarkable, scaling model based on an *effective* elastic modulus (the secant modulus at the stability limit) gives successful predictions for the onset speed of the crack instability.

One may ask, why is it that scaling laws are of such distinguished importance? The answer is that scaling laws never appear by accident. (G.I. Barenblatt, *Scaling*. Cambridge University Press, 2003)

6.1 Introduction

In 1951, Yoffe [1] made the physically intuitive suggestion that mode I crack growth occurs in the direction of *maximum asymptotic hoop stress* and found the crack speed for the onset for branching to be about 70% of the Rayleigh wave speed c_R [2, 3]. This high speed is rarely observed in experiment [4, 5]. An obvious shortcoming in Yoffe's analysis is the assumption of a constant linear elastic response for all deformations.

We show by atomistic simulation that hyperelasticity, the nonlinear elasticity of crystal deformation, plays a governing role in the instability dynamics of brittle fracture: i.e. hyperelasticity significantly influences the critical speed at which a crack deviates from forward motion. However, we find that the *local* hyperelasticity in the crack tip's neighbourhood does not quantitatively explain our simulation results. We discover an *effective* hyperelasticity modulus that describes a nonlocal (global) elastic response and yields a universal scaling of the instability phenomenon.

Our simulation model is based on a generalized bilinear force law composed of two spring constants, one associated with small deformations (k_1, $r \leq r_{on}$)

Springer Proceedings in Physics, Volume 105
Computer Simulation Studies in Condensed-Matter Physics XVIII
Eds.: D.P. Landau, S.P. Lewis and H.-B. Schüttler

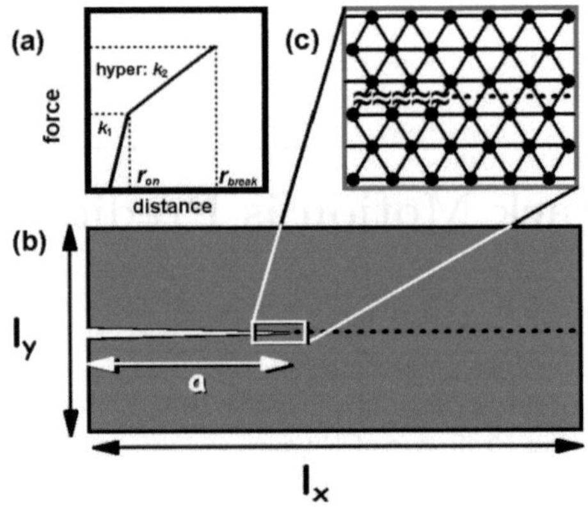

Fig. 6.1. (**a**) The bilinear force is composed of two spring constants, one associated with small deformations (k_1, $r \leq r_{\rm on}$.) and the other associated with large deformations (k_2, $r > r_{\rm on}$). (**b**) The two-dimensional simulation geometry is shown, the slab size being given by l_x, l_y and the crack extension by a. (**c**) The crack initially propagates in a triangular hexagonal lattice with nearest neighbour distance $r_0 = 2^{1/6}$ along a cleavage plane with lowest surface fracture energy (denoted by *dotted line*)

and the other associated with large deformations (k_2, $r > r_{\rm on}$). This is shown in Fig. 6.1a. This serves as a useful model, allowing us to investigate the generic effects of hyperelasticity by changing the relative magnitude ($\alpha = k_2/k_1$) and transition distance ($r_{\rm on}$) of the potential. This bilinear force has been used to investigate the dynamics of cracks constrained to remain straight [6].

Depending on the choice of interatomic potential parameters for our model solid, we define an *effective* elastic modulus which has a remarkable relationship to the crack's instability dynamics. We discover that the crack's speed at its instability onset collapses onto a universal linear dependence with this new modulus. This finding allows for successful predictions for the crack's instability speed in nonlinear materials. Furthermore, we note that simulation of a strictly linear solid agrees with Yoffe's solution. We suggest the possibility that Yoffe's picture of the dynamic instability in brittle fracture may be valid provided the variable elasticity of the solid is properly represented. Such a suggestion requires theoretical verification.

6.2 Modelling

We consider the propagation of a crack in a two-dimensional crystal geometry shown in Fig. 6.1b. The slab lengths are given by l_x and l_y, respectively. The crack propagates in the x-direction, and its extension is denoted by a. The slab is loaded in mode I in the y-direction with a constant strain rate equal to 0.00001 (all quantities given are in dimensionless units). A slit of length $a = 200$ is cut midway through the slab as an initial, atomically sharp crack.

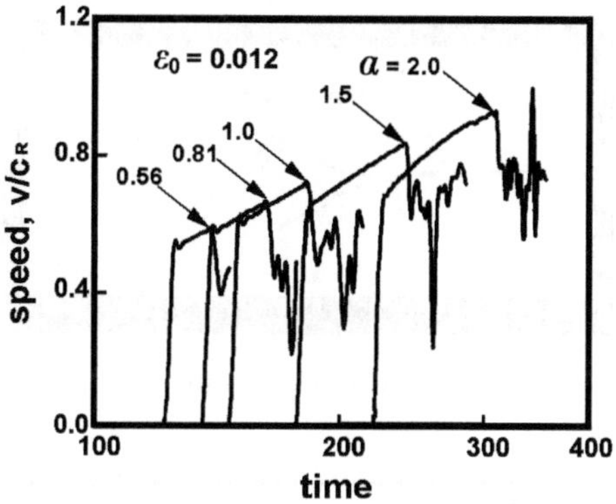

Fig. 6.2. A comparison of crack speed histories is shown for the bilinear solid with different k_2/k_1 and a particular transition distance $r_{\text{on}} = 1.136$ expressed as $\varepsilon_0 = (r_{\text{on}}/r_0) - 1 = 0.012$. The dynamic crack instabilities for the various $\alpha = k_2/k_1$ are indicated by the *arrows*

The crack initially propagates in a triangular hexagonal lattice with nearest neighbour distance $r_0 = 2^{1/6} \approx 1.12246$ along a cleavage plane with lowest surface fracture energy (Fig. 6.1c). We assume that atomic bonds snap at a critical atomic snapping distance $4_{\text{break}} = 1.17$. The slab is initialized at zero temperature.

We adopt a bilinear, interatomic force law composed of two spring constants $k_1 = 72/\sqrt[3]{3} \approx 57.14$ and $k_2 = \alpha k_1$, $\alpha = 0.5625, 0.81, 1.0, 1.5$ and 2.0. The spring constant k_1 is associated with displacements from the equilibrium distance r_0, and the second spring constant k_2 is associated with large bond stretching for $r > r_{\text{on}}$. For each k_2, the values for r_{on} are taken to be 1.1257, 1.1290, 1.1360, 1.1375, 1.1449 and 1.1550, respectively. There are three distinct wave velocities in elastic solid: the longitudinal wave speed $c_l = \sqrt{3\mu/\rho}$, the shear wave speed $c_5 = \sqrt{\mu/\rho}$ and the Rayleigh wave speed $c_{\text{R}} \approx 0.9225 c_5$, with the density $\rho = 2/\sqrt{3}/\sqrt[3]{2} \approx 0.9165$ for atomic mass $m = 1$. For small deformation, the shear modulus $\mu = 24.8$.

6.3 Results

Figure 6.2 shows a comparison of crack speed histories for the bilinear solid with different k_2/k_1 and a particular transition distance $r_{\text{on}} = 1.136$ expressed as $\varepsilon_0 = (r_{\text{on}}/r_0) - 1 = 0.012$. The histories are similar for $\varepsilon_0 = 0.003, 0.006, 0.014, 0.020$ and 0.029, corresponding to the respective r_{on}. The dynamic crack instabilities for the various $\alpha = k_2/k_1$ are associated with the precipitous drops in crack speed, as indicated by the arrows, and are a consequence of the crack deviating from straight line motion (see Fig. 6.3). The crack speed at the onset of erratic motion is defined as the instability speed. Figure 6.4

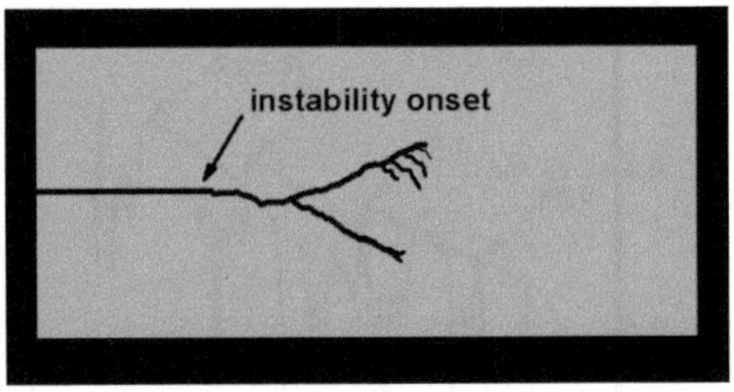

Fig. 6.3. A picture of a crack is shown at a significant time beyond the onset of the instability

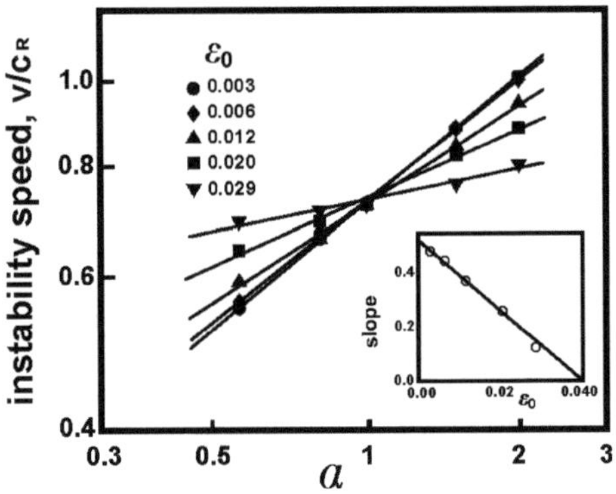

Fig. 6.4. A log–log plot of the instability speed as a function of $\alpha = k_2/k_1$ is presented for various $\varepsilon_0 = (r_{\text{on}}/r0) - 1$

presents a log–log plot of the instability speed as a function of $\alpha = k_2/k_1$ for various $\varepsilon_0 = (r_{\text{on}}/r_0) - 1$. For each ε_0, the dependence is essentially linear, the slope approaching one-half for ε_0 tending to zero. This trend is required since $k_2/k_1 = 1$ for $\varepsilon_0 = 0$ and the solid is strictly linear with a spring constant equal to k_2. Therefore, the instability speed will have a trivial square-root dependence on the spring constant k_2 when normalized by k_1. The other limit is $r_{\text{on}} = r_{\text{break}}$ ($\varepsilon_0 = 0.04235$). In this limit, the bilinear force law is simply the linear force with spring constant k_1.

Figure 6.5 defines, graphically, our choice for an *effective* spring constant k_{eff} of the bilinear force law. It is simply the slope of the vector sum of the maximum piecewise linear forces defined by the bilinear force law as shown in Fig. 6.5; i.e.

$$k_{\text{eff}} = k_1 \left[\frac{(r_{\text{on}} - r_0)}{(r_{\text{break}} - r_0)} \right] + k_2 \left[\frac{(r_{\text{break}} - r_{\text{on}})}{(r_{\text{break}} - r_0)} \right] .$$

The elastic modulus associated with this effective spring constant is the secant modulus at the mechanical stability limit.

By plotting in Fig. 6.6 the instability speed as a function of $\alpha_{\text{eff}} = k_{\text{eff}}/k_1$, we see a remarkable collapse of the data from Fig. 6.4 onto a common straight

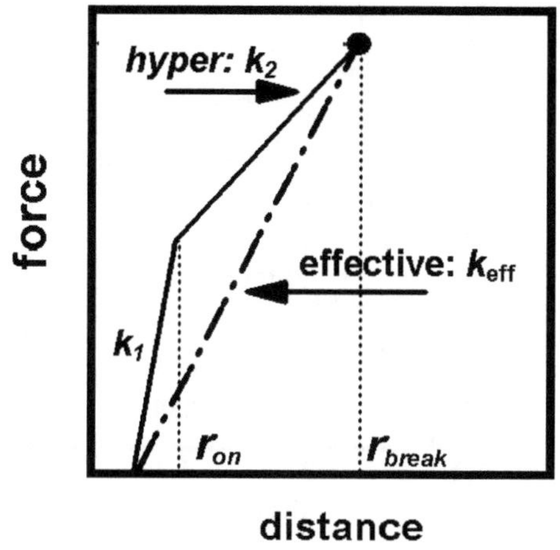

Fig. 6.5. The *effective* spring constant k_{eff} is defined graphically for the bilinear force

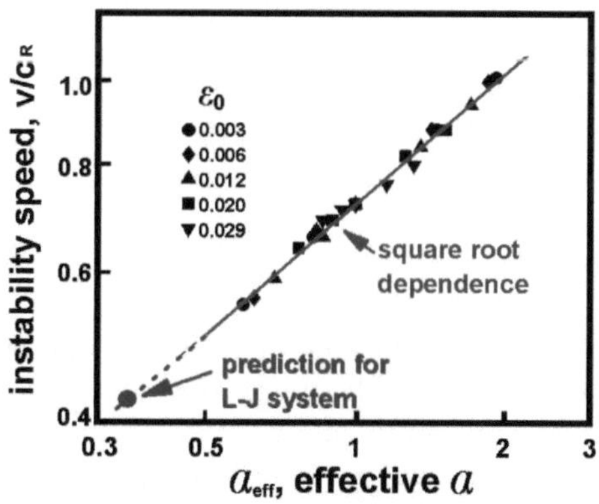

Fig. 6.6. The instability speed is presented as a function of $\alpha_{\mathrm{eff}} = k_{\mathrm{eff}}/k_1$

line with slope equal to one-half. Hence, for determining the instability speed of a dynamic brittle crack, this finding allows one to model the bilinear material as a linear solid with the effective spring constant just described.

We apply this concept of an effective spring constant to a continuous interatomic potential: in particular, to the Lennard–Jones 12:6 potential. Our construction of $\alpha_{\mathrm{eff}} = k_{\mathrm{eff}}/k_1$ is shown in Fig. 6.7, and it should be self-explanatory. Consistent with our geometrical construction of k_{eff} for the bilinear force, we adopt the slope for the line passing through the LJ force curve at its equilibrium and its maximum as the effective spring constant. We have defined the spring constant with an L–J subscript for sake of clarity. It is a big jump of faith to apply this simple construction to a continuous potential, but the prediction is quite sensible. In Fig. 6.6, we note that it predicts an

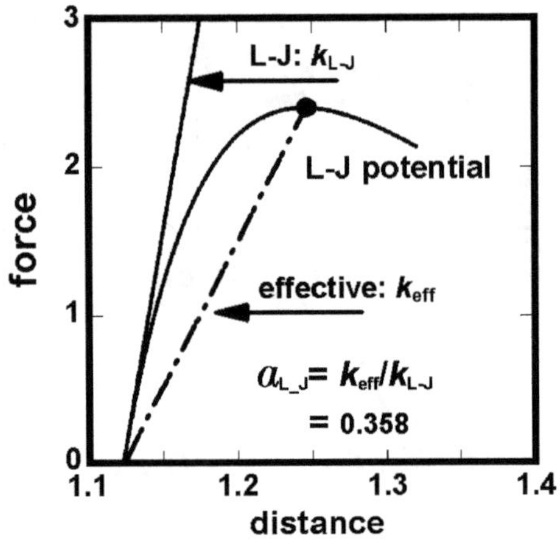

Fig. 6.7. The construction of $\alpha_{\text{eff}} = k_{\text{eff}}/k_1$ is shown

instability speed of ≈ 0.42, not much different from speed of 0.36 observed in computer simulations [7, 8].

For a simple linear solid, the instability speed is 0.73. For a nonlinear solid, the instability speed is $0.73\sqrt{(k_{\text{eff}}/k_1)}$.

6.4 Discussion

Recent atomistic fracture simulations of a linear solid have shown remarkable agreement between the virial hoop stress of statistical mechanics and the asymptotic solution of continuum mechanics [9]. In that study, the atomic crack was constrained to propagate along a straight path so that such a comparison could be made for crack speeds beyond the instability speed. The main result of that study was that the hoop stress near a crack tip using continuum theory is reproduced well in the atomistic simulations. Most important, the Yoffe prediction that the hoop stress field becomes bimodal above 73% of the Rayleigh speed is found in MD simulation results as well.

In Fig. 6.4, we note that the instability speed for the strictly linear solid ($\alpha = 1$) is equal to 73% of the Rayleigh speed, in agreement the Yoffe prediction. This suggests that Yoffe's picture of the dynamic instability in brittle fracture may be valid. It is only necessary to replace the elastic modulus for small deformation with an *effective* elastic modulus (the secant modulus) described in this study, giving successful predictions for the onset speed of the crack instability for nonlinear materials. The challenge is to theoretically demonstrate the simulation finding, i.e., the simple scaling relation. The mathematical study could be based on a Yoffe-type analysis assuming a piecewise linear elasticity consistent with the bilinear elastic modulus defined our model force law. Such an analysis should answer whether the Yoffe picture is the basis of the instability phenomenon. Of course, the successful theory must explain our remarkable scaling law.

Finally, our finding suggests that theories that are based strictly on the *local* elastic properties in the neighbourhood of the crack tip (e.g., [10, 11]) would be incomplete since our study shows that materials failure depends on an effective *global* modulus of the solid (the secant modulus at the stability limit).

References

1. E.H. Yoffe, Phil. Mag. **42**, 739 (1951)
2. B. Broberg, *Cracks and Fracture.* (Academic Press, San Diego 1999)
3. L.B. Freund, *Dynamic Fracture Mechanics.* (Cambridge University Press, Cambridge 1998)
4. J. Fineberg, S.P. Gross, M. Marder, H.L. Swinney, Phys. Rev. Lett. **67**, 141 (1991)
5. K. Ravi-Chandar, Int. J. Fract. **90**, 83 (1998)
6. M. Buehler, F.F. Abraham, H. Gao, Nature **426**, 141 (2003)
7. F.F. Abraham, D. Brodbeck, R.A. Rafey, W.E. Rudge, Phys. Rev. Lett. **73**, 272 (1994)
8. F.F. Abraham, D. Brodbeck, W.E. Rudge, X. Xu, J. Mech. Phys. Solids **45**, 1595 (1997)
9. M. Buehler, F.F. Abraham, H. Gao, *Stress and energy flow field near a rapidly propagating mode I crack.* SIAM Multiscale Modeling and Simulation (to appear 2004)
10. F.F. Abraham, Phys. Rev. Lett. **77**, 869 (1996)
11. H. Gao, J. Mech. Phys. Solids **44**, 1453 (1996)

7

First Principles Theory of Nano-Materials, Spintronic Materials, and Nano-Spintronic Materials

L. Kronik

Department of Materials and Interfaces, Weizmann Institute of Science,
Rehovoth 76100, Israel

Abstract. Two of the most exciting recent developments in semiconductor science and technology are the advent of semiconductor nanocrystals and of dilute magnetic semiconductors. Here, the significance of first principles theory for elucidating the properties of these material classes is explained and numerous examples are given. The examples include studies of the quantum size effect in semiconductor nanocrystals, studies of the spin-polarized electronic structure in bulk dilute magnetic semiconductors, and studies of size-dependent electronic and magnetic properties in dilute magnetic semiconducting nanocrystals.

7.1 Introduction

Two of the most exciting recent developments in semiconductor science and technology are the advent of semiconductor nanocrystals and dilute magnetic semiconductors (DMSs). Semiconductor nanocrystals are surface-passivated, nanometer sized (typically spherical) objects possessing a bulk-like crystalline structure. They have been studied intensively because of their unusual electronic and optical properties, which may differ fundamentally from those of the corresponding bulk material. Of particular appeal is that the properties of nanocrystals can be radically altered, while maintaining their chemical composition, simply by changing their size and/or shape [1]. A striking maneifestation of that is the "quantum size effect", where the optical absorption threshold of, e.g., CdSe nanocrystals can be tuned from the infrared to the ultraviolet, across the entire visible spectrum, merely by changing the nanocrystal size [2].

DMSs are semiconductors to which (typically several percent of) a magnetic impurity has been intentionally introduced [3]. They have attracted considerable attention because they form a class of materials with both semiconducting and magnetic (in some cases even ferromagnetic) properties. DMSs also hold the promise of using electron spin, in addition to its charge, for

Springer Proceedings in Physics, Volume 105
Computer Simulation Studies in Condensed-Matter Physics XVIII
Eds.: D.P. Landau, S.P. Lewis and H.-B. Schüttler

creating a new class of spintronic semiconductor devices with unprecedented functionality [4].

Both groups of materials present fundamental challenges to theory because the rich phenomena they display is often outside the realm of traditional models. For example, in nanocrystals the quantum size effect can be explained qualitatively by effective mass considerations. However, the effective mass approximation breaks down in the nano-regime and quantitative predictions based on it fail [1]. As a second example, the mechanisms controlling the magnetic order in DMSs may be very different than those found in traditional magnetic materials, so that qualitative arguments are often controversial [5].

The need to tread where "traditional" models do not necessarily apply naturally leads to first principles theoretical calculations, i.e., calculations based on nothing other than the atomic number of the constituent species and the laws of quantum physics as input. Such calculations are inherently not affected by the "bias" which may be present in phenomenological models and are therefore not subject to the pitfalls of the latter. Here, I will illustrate this first principles approach by reviewing some recent calculations of size-dependent optical effects in nanocrystals, the spin-polarized electronic structure in DMSs, and the size-dependence of the spin-polarized electronic structure in nanocrystalline DMSs. These calculations, based on ab initio pseudopotentials within density functional theory, are intended to provide explanations and/or predictions of the observed phenomena, as well as to aid in developing a qualitative and phenomenological understanding of the pertinent physical mechanisms.

7.2 Methodology

In this section, I briefly describe the computational approaches used for obtaining the results presented in the following sections. I deliberately avoid a rigorous mathematical derivation (which can be found in the references provided below) and instead focus on the key physical ideas and concepts.

One of the most popular approaches for large-scale first principles computations is the use of ab initio pseudopotentials within density functional theory (DFT) [6]. This approach has been used successfully for predicting, e.g., mechanical, chemical, and electronic properties of many classes of solids, liquids, molecules, and more.

The pseudopotential-density functional approach weds two physically reasonable approximations that result in a dramatic reduction of computational complexity. Within density functional theory, the original N-electron problem is mapped into an effective one-electron equation (known as the Kohn–Sham equation), where all non-classical electron interactions (namely, exchange and correlation) are subsumed into an additive one-electron potential that is a functional of the charge density [7]. While this mapping is formally exact, it is approximate in practice because the exact functional is unknown. A simple approximate functional, which is used consistently throughout this work, is the local density approximation (LDA). In LDA, the exchange-correlation functional is taken to be a *local* function of the charge density. Mathematically, DFT converts a differential eigenvalue problem in 3N spatial coordinates (which is practically intractable for all but the smallest systems) into a standard eigenvalue problem with only 3 spatial coordinates.

In the pseudopotential approximation, only valence electrons (the ones forming chemical bonds) are treated explicitly. Core electrons are suppressed by replacing the true atomic potential with an effective "pseudopotential" that takes the effect of core electrons into account. This facilitates DFT calculations in two ways: First, the pseudopotential is a smooth, slowly-varying potential, whereas the true atomic potential is rapidly varying and has a $\sim 1/r$ singularity at the nuclear position. This makes the attainment of a converged numerical solution significantly easier. Second, treating only valence electrons reduces the number of eigenvalues that need to be found, often by as much as an order of magnitude. Here, we use Troullier–Martins pseudopotentials [8]. In this approach smooth, atomic "pseudo-wave-functions" that differ from the original ones only inside a given core radius are constructed. Thus, the chemical binding properties associated with the wave function values outside the core are maintained. The pseudopotential is then constructed so as to yield the smooth pseudo-wave-function.

For periodic systems, the Kohn–Sham equations can then be solved relatively easily by expanding potentials and functions in terms of a planewave basis [6]. This is because the planewave approach inherently accounts for the system periodicity and because the smooth nature of the pseudo-wave-functions and pseudopotentials facilitates rapid convergence with basis size.

The same approach can be used for non-periodic structures, including the nanocrystalline systems discussed below, with the help of a "supercell", where the system of interest is artificially replicated. This approach, however, also has significant drawbacks. A paramount problem is the need to eliminate (or at least minimize) spurious interactions between adjacent supercell replicas. Compromises based on practical computational limitations may cause significant errors in interpretation of the results [9]. Another problem is the treatment of charged systems. If each supercell replica is charged, the system has an infinite overall charge and its energy diverges. This is remedied if a uniform compensating charge density is inserted, but this may again cause the computed system to differ from the true one in non-trivial ways. Finally, even if these problems do not arise, the heavy use of fast Fourier transforms made by planewave codes makes a massively parallel implementation non-trivial.

An approach which alleviates the above disadvantages is the use of a real-space grid instead of an explicit basis set [10], with the kinetic energy (Laplacian) operator expressed as a high-order finite difference involving many neighbouring points. This typically results in a sparse matrix representation that can be solved using highly effective, massively parallel algorithms. This approach was used for all the nanocrystalline systems discussed below.

Finally, the above described formalism is insufficient for describing electronic excitations in general and optical properties in particular. This is because traditional, time-independent DFT is *inherently* a ground-state theory [7]. A generalized, time-dependent DFT (TDDFT) formalism that allows for computations of excited state properties has been developed [11] and applied successfully to a wide range of finite systems [12]. Briefly, we use a frequency-domain formulation of TDDFT within the local density approximation (TDLDA) [11], where poles and residues of the frequency-dependent polarizability are related to optical transition energies and oscillator strengths, respectively.

7.3 The Quantum Size Effect: Silicon Versus Germanium Nanocrystals

The optical gap of bulk Ge, $\sim 0.7\,\text{eV}$, is significantly smaller than that of Si, $\sim 1.1\,\text{eV}$. Takagahara and Takeda [13], using an effective mass theory, predicted that because the effective mass of charge carriers in Ge is much smaller than in Si, surface-passivated Ge nanocrystallites smaller than some critical radius would exhibit an optical gap *larger* than that of Si. Later, Hill et al., using a tight binding approach, reached a qualitatively similar conclusion [14], but revised the value of the critical radius to a smaller one. This predicted "crossover" between the optical gaps of Si and Ge has since been questioned. Reboredo and Zunger [15], using an empirical pseudopotential approach, argued that there is no clear crossing of optical gap vs size. More recently, Weissker et al. [16] estimated the optical gaps of Si and Ge nanocrystals using the delta self-consistent field (ΔSCF) method and did not find a clear crossover either.

Clearly, such a situation calls for first principles calculations that do not employ any parametrization and do not require any experimental input[1]. We have therefore performed TDLDA calculations for Si and Ge nanocrystals [17]. Approximately spherical hydrogen-passivated Si and Ge nanocrystals were considered, where all Si or Ge atoms were assumed to be located at their ideal bulk positions, and hydrogen atoms were placed so as to passivate every Si or Ge dangling bond. The smallest nanocrystals thus constructed were SiH_4 or GeH_4, which are simply the silicon or germane molecule, respectively, and the largest were $Si_{147}H_{100}$ and $Ge_{147}H_{100}$.

The results, together with those of the above-mentioned previous studies, are shown in Fig. 7.1. Clearly, we do not find a crossover between the Si and Ge

[1] ΔSCF actually is a first principles method in the sense that it uses no empirical input. However, it is not rigorously anchored within DFT because of the use of a partially filled orbital that is not the highest occupied state. This may result in uncotrolled errors.

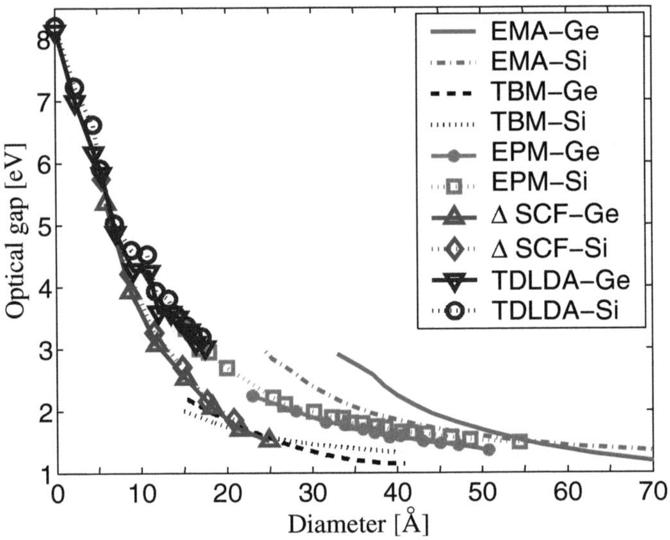

Fig. 7.1. Optical gap of Ge and Si nanocrystals as a function of diameter, computed using TDLDA [17], effective mass approximation [13], tight binding [14], empirical pseudopotentials [15], and ΔSCF [16]. (After Nesher et al. [17]

optical gap curves for any crystallite size, although for small enough crystallites the curves closely track each other. As explained above, our nanocrystals are taken such that the limiting case "zero diameter" nanocrystal is silane or germane. Figure 7.1 clearly predicts that the optical gap order is preserved even for this molecular limit. Indeed, experimentally the optical gap of germane is smaller than that of silane −8.3 vs. 8.8 eV, respectively [18].

Qualitatively, our results provide first principles support for the conclusions of Reboredo and Zunger [15] and of Weissker et al. [16] and disagree with the earlier predictions of Takagahara and Takeda [13] and of Hill et al. [14]. Quantitatively, however, our results agree with those obtained from semiempirical pseudopotentials [15], but differ from the ΔSCF results [16], the latter being similar to the TDLDA ones for the smallest nanocrystals, but lower in energy for the larger ones. These findings clearly illustrate the need for first principles theory where empirical calculations are controversial.

7.4 Dilute Magnetic Semiconductors: MnGaAs and MnGaN

In recent years, by far the most studied and used DMS was $Mn_xGa_{1-x}As$, a material inherently compatible with the relatively mature GaAs technology, and from which spin-polarized charge injection to GaAs was successfully demonstrated [19]. This material also has two main drawbacks: First,

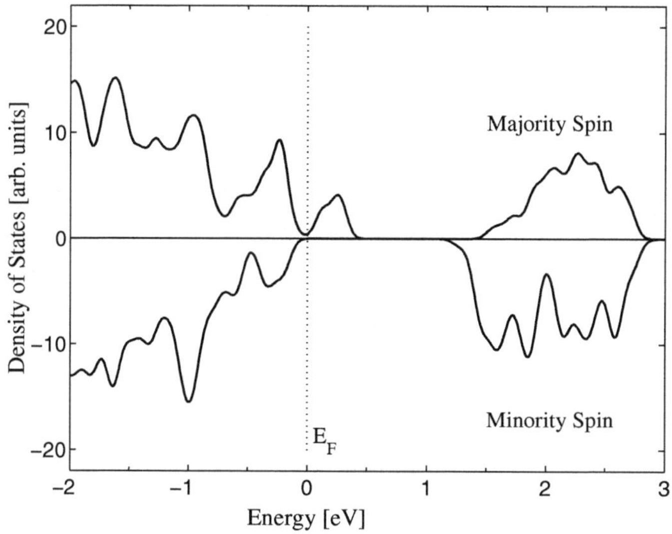

Fig. 7.2. Spin-polarized density of states for $Mn_{0.063}Ga_{0.937}As$. (After Jain et al. [21]

its Curie temperature is low ($\sim 110\,K$), hindering room temperature operation. Second, because of a low miscibility limit of Mn in GaAs, $Mn_xGa_{1-x}As$ films are usually grown using low temperature molecular beam epitaxy (MBE) and therefore exhibit a limited electronic quality. More recently, $Mn_xGa_{1-x}N$ has attracted considerable interest. The Curie temperature of this material is higher [20], and Mn is more miscible in GaN than in GaAs. However, the technology of this material (or, indeed, of GaN in general) is considerably less developed.

While both alloys are ferromagnetic, this is not sufficient for a successful operation of spintronic devices. Support of spin-polarized transport so that spin-polarized charge carriers may be injected into a non-magnetic semiconductor is also required. With experimental analyses difficult due to the quality of the material, we examined theoretically the limits to spin-polarized transport in both $Mn_xGa_{1-x}As$ and $Mn_xGa_{1-x}N$, with a realistic $x = 0.063$, by computing their spin-polarized electronic structure from first principles [21, 22].

Calculations for the density of states of $Mn_xGa_{1-x}As$ and $Mn_xGa_{1-x}N$ are presented in Figs. 7.2 and 7.3. Clearly, both majority and minority spin components of either alloy display a band-gap, indicating that the introduction of Mn did not destroy the semiconducting nature of the material. (The band-gap of both materials is underestimated with respect to the experimental value – a well-known effect of density functional theory, resulting from it being a ground state theory not suited for the computation of excited state properties, as mentioned above). The most striking result of the introduction

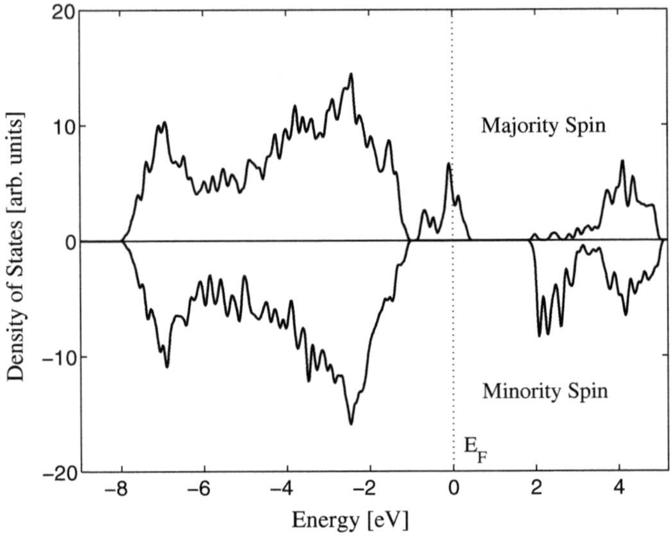

Fig. 7.3. Spin-polarized density of states for $Mn_{0.063}Ga_{0.937}N$. (After Kronik et al. [22]

of Mn is the formation of a Mn-related, ideally spin-polarized features, with a width of $\sim 0.5\,eV$ in $Mn_{0.063}Ga_{0.937}As$ and $\sim 1.5\,eV$ in $Mn_{0.063}Ga_{0.937}N$. Orbital analysis of the spin-polarized features indicates that they are due to significant hybridization of the Mn 3d states with As 4p or N 2p states.

Importantly, the Fermi level is within the spin-polarized feature in both cases. This means that both materials are half-metallic, i.e., their majority spin density of states is metallic, but their minority spin density of states is semiconducting. Because transported charge carriers come from the vicinity of the Fermi level, they will have a well-defined (majority) spin. Therefore, the theoretical limit to spin-polarized transport is an ideal 100%. It is important to note that in practice the position of the Fermi level is determined by "unintentional doping" of defects and residual impurities. If these effects are significant, "Fermi level engineering", performed by adding a different donor or acceptor species (in trace amounts, such that the band structure is not affected), may be required for bringing the Fermi level to within the impurity band. Such engineering needs be done with care as it would also affect the free carrier density and therefore the Curie temperature.

The spin-polarization of $Mn_{0.063}Ga_{0.937}As$ and $Mn_{0.063}Ga_{0.937}N$ has a different physical origin. In the former, the spin-polarized feature clearly results from a Mn-induced spin-splitting of the valence band. In the latter, the valence band remains essentially unpolarized. Instead, an independent, spin-polarized impurity band arises. Physically, the sharp difference in the nature of the Mn-related spin-polarized features is due to the nature of the Mn impurity in the host material. In GaAs, Mn is a shallow acceptor, situated $\sim 0.1\,eV$ above the valence band maximum in the isolated impurity limit [23]. It therefore hybridizes primarily with the valence band and completely merges with it for a few percent of Mn. In contrast, Mn is a deep acceptor in GaN – forming a level that is well over an eV above the valence band maximum [24]. Its interaction with the valence band is therefore much smaller, and the introduction of Mn barely polarizes the valence band. For $x = 0.063$, the impurity band does not hybridize to an extent sufficient for merging with the valence band.

Since the above-described theoretical predictions have been first made, the half-metallic nature of MnGaAs has been confirmed experimentally using time-resolved magneto-optical measurements [25]. For MnGaN, such direct proof is not available yet. However, recent experiments did confirm the existence of the spin-polarized impurity band by showing that depletion of carriers from it may eliminate the ferromagnetic order [26]. These findings show the importance of first principles predictions in guiding experimental efforts.

7.5 Dilute Magnetic Semiconducting Nanocrystals: Ge:Mn, GaAs:Mn, and ZnSe:Mn

Combining the lessons of the two previous sections, we expect nano-crystalline DMSs to exhibit intriguing magnetic properties, which may be different than those of the bulk because quantum confinement is known to enhance spin–spin interactions [1]. As such, it is important to understand the role of dimensionality in shaping the spin-polarized electronic structure of nanocrystalline DMS [4]. Quantum confinement may also have technological implications because magnetic dots have been suggested for use in quantum computation [27] and because semiconducting nanoparticles can be connected via conjugated molecules [28], providing an effective means for spin communication between dots. Despite the importance of magnetic dots, no systematic study of the electronic structure of Mn-containing nanocrystalline DMS from first principles has been undertaken. Moreover, we are aware of only one paper where size effects, on a particular DMS ($Mn_xGa_{1-x}As$, with x fixed as a function of size), have been examined theoretically [29].

For elucidating the potential of nanocrystalline DMSs theoretically, we studied the electronic structure and magnetic properties of Mn-containing Ge, GaAs, and ZnSe nanoparticles. This choice of semiconducting materials was motivated by several considerations. All three are well-known semiconductors, which are prototypical of group IV, III–V, and II–VI semiconductors. The elements comprising these semiconductors are found in the same row of the periodic table and these semiconductors have the same (zincblende / diamond) crystalline structure, making an identification of chemical trends easier. Also, Mn-based bulk DMSs have been successfully synthesized in all three cases [3, 30, 31]. Finally, a successful synthesis of high-quality Mn-containing ZnSe nanocrystals has been reported [32].

For the theoretical studies, we passivated the Ge nanocrystal surface using hydrogens as before, whereas the nanocrystal surfaces of GaAs and ZnSe were passivated using fictitious, hydrogen-like atoms with fractional charge [33]. Because magnetic circular dichroism experiments have suggested that each nanocrystal contains, on average, only one Mn atom [32], we started by placing one Mn atom in the nanocrystal. A four-fold coordinated Mn atom was placed in the center of the Ge nanocrystal. In GaAs and ZnSe, the Mn atom was substituted on a cation site in the center of the nanocrystal. We considered four Mn-doped nanocrystals: $X_9\mathrm{Mn}Y_{10}$, $X_{18}\mathrm{Mn}Y_{19}$, $X_{40}\mathrm{Mn}Y_{41}$, and $X_{64}\mathrm{Mn}Y_{65}$, where $X =$ Ge, Ga, or Zn and $Y =$ Ge, As, or Se, respectively (the passivation atoms are implicit).

Figure 7.4 shows the spin-polarized energy levels for the 82-atom system for all three materials systems [34]. In all three cases, the presence of Mn introduces gap states that are derived from Mn $3d$ states and reflect the splitting of the $3d$ states by the T_d crystal field to an e doublet and a t_2 triplet. An analysis of the corresponding wave functions shows that the valence band edge is comprised mainly of anion p states. The majority spin e levels are fully occupied, hybridized with anion p states, and located right below ($\sim 0.05\,\mathrm{eV}$) the valence band edge in all three materials. The majority t_2 levels are characterized by a fairly large hybridization with the four neighboring p orbitals and the corresponding charge density is highly localized on the $\mathrm{Mn}Y_4$ complex. They are singly, doubly, and triply (fully) occupied for Ge:Mn, GaAs:Mn, and ZnSe:Mn, respectively.

Both bulk Ge:Mn [35] and the above-discussed GaAs:Mn [21] are half-metallic. However, bulk ZnSe:Mn was found to be semiconducting [36], i.e., the Fermi level was in the forbidden gap for both spin channels. The origins of this behavior are apparent here. Figure 7.4 shows that the e and t_2 minority spin levels are empty, so that the minority spin retains its semiconducting nature even in the presence of the Mn impurity in all three cases. However, the highest occupied orbital (the t_2 level) of the majority spin is partially occupied for Ge:Mn and GaAs:Mn, but is fully occupied for ZnSe:Mn. This configuration is in agreement with a "half-metallic" nature of majority spin electrons for Ge and GaAs, and in agreement with the semiconducting nature for ZnSe. Thus, the "origins" of the half-metallic behavior of the bulk phase (or lack thereof) are clear already at the nanocrystal level.

A consequence of this electronic structure configuration is that the introduction of the Mn impurity does not change the number of minority spin *occupied* states. Therefore, the introduction of the 5 Mn d electrons results in a net magnetic moment of $3\,\mu_B$, $4\,\mu_B$, and $5\,\mu_B$ for Ge, GaAs, and ZnSe, respectively. This is in agreement with the magnetic moment of $5\,\mu_B$ for the free Mn atom being modified by the doubly ionized acceptor, singly ionized acceptor, and isoelectronic nature of Mn in Ge, GaAs, and ZnSe, respectively, with the holes associated with the acceptor level localized in the t_2 orbitals [37].

Investigations of the electronic structure as a function of nanocrystal size reveal that the Mn-related states are practically size-independent because

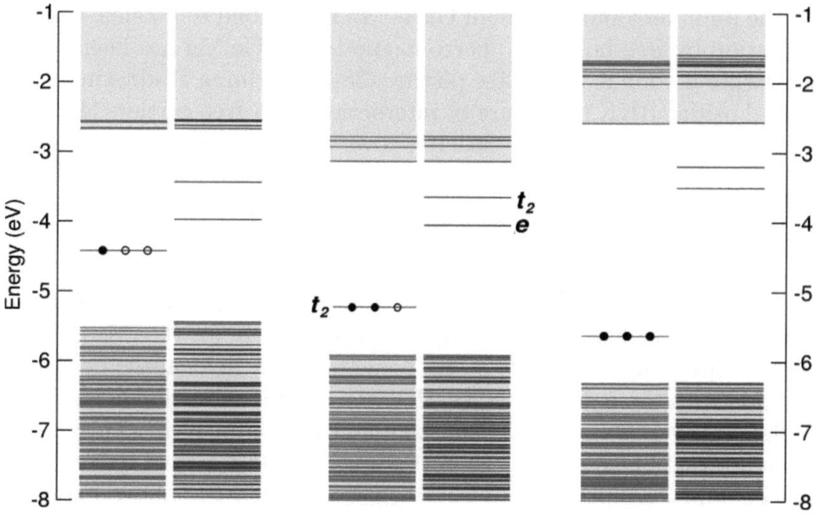

Fig. 7.4. Spin-polarized electronic structure for passivated (**a**) $Ge_{81}Mn$, (**b**) $Ga_{40}MnAs_{41}$, and (**c**) $Zn_{40}MnSe_{41}$ nanocrystals. The e and t_2 levels are doubly and triply degenerate, respectively. The Fermi level is located at the majority spin t_2 levels in all cases. Filled and empty circles denote electrons and holes, respectively. (After Huang et al. [34]

of the localized nature of the associated orbitals. The significantly different quantum size effects experienced by localized and delocalized orbitals have profound implications for the resulting electronic structure and magnetic properties. For example, we have already pointed out that Mn is a shallow acceptor in bulk GaAs. However, for the 130-atom nanocrystal, the Mn t_2 states are ~ 0.6 eV above the "host HOMO", because the delocalized occupied orbitals are pushed down in energy more rapidly then the Mn states with decreasing size [29]. This means that quantum size effects transform Mn from being a shallow acceptor to being a deep acceptor.

To investigate the effect of size on impurity–impurity spin interactions, we introduced two Mn atoms into each 82-atom nanocrystal. First, we considered the case where both Mn atoms are placed on adjacent cation sites, bridged through an anion (in Ge one bridging Ge atom was placed between the two Mn atomss). We compared the total energy differences between the ferromagnetic (FM, parallel spins) and antiferromagnetic (AFM, anti-parallel spins) configurations. We found the FM structure to be more stable by 0.42 eV and 0.38 eV for Ge and GaAs, respectively, but less stable by 0.14 eV for ZnSe. This is in agreement with bulk results, where the stable phase of Ge:Mn and GaAs:Mn is predicted to be the FM one, whereas in ZnSe:Mn the stable phase is the AFM one. Also in agreement with bulk studies [38] is the fact that if both Mn atoms are nearest neighbors (i.e., they form a dimer) in Ge:Mn, the AFM phase is always more stable.

In the bulk, ferromagnetism in GaAs:Mn is ascribed to a Zener-like picture of mediation by free holes [39]. Ferromagnetism in Ge:Mn has been explained either within a similar Zener-like picture [38] or within a Rudderman–Kittel–Kasuya–Yosida (RKKY) picture of interaction with free carriers [40]. For the nanocrystalline case, Fig. 7.4 clearly shows the absence of either free holes in the valence band or a metal-like presence of free carriers, which precludes either mechanism (the electronic structure diagram of the FM nanocrystal containing 2 Mn atom is qualitatively similar to the 1 Mn atom case). We conclude that the FM interaction in nanocrystalline Ge:Mn and GaAs:Mn is *different from that of the bulk* and is stabilized via a double exchange mechanism involving localized holes as suggested previously for (Ga,Mn)N [5]. This picture is consistent with the t_2 states being partially delocalized because of an interaction with the polarized holes. However, hole localization in nanocrystals is much greater than in the bulk due to quantum confinement. This makes the Mn–Mn interaction essentially a short-range one. The lack of ferromagnetism in the ZnSe:Mn nanocrystals is also consistent with the above explanation. In ZnSe:Mn, the t_2 states are fully occupied (see Fig. 7.4) and there are no holes to couple to. This shows how first principles calculations can be used to predict and rationalize quantum size effects on magentic properties.

7.6 Conclusion

In this paper, we have shown how first principles calculations can serve as a valuable tool for prediction of properties and elucidation of physical mechanisms in semiconductor nanocrystals, dilute magnetic semicondcutors, and dilute magnetic semicondcuting nanocrystals. They thus serve as an important bridge between experiment and formal theory.

I wish to thank the Delta Career Development Chair, the Minerva Foundation, and PHOREMOST – the European network of excellence on nanophotonics, for financial support.

References

1. A.L. Efros, M. Rosen: Annu. Rev. Mater. Sci. **30**, 475 (2000)
2. B. Murray, D.J. Norris, M.G. Bawendi: J. Am. Chem. Soc. **115**, 8706 (1993)
3. H. Ohno: Science **281**, 51 (1998)
4. S.A. Wolf, D.D. Awschalom, R.A. Buhrman, J.M. Daughton, S. von Molnar, M.L. Roukes, A.Y. Chtchelkanova, D.M. Treger: Science **294**, 1488 (2001)
5. T. Graf, S.T.B. Goennenwein, M.S. Brandt: Phys. Stat. Sol. (b) **239**, 277 (2003)
6. J.R. Chelikowsky, M.L. Cohen, in *Handbook on Semiconductors*. 2nd edn., ed. by T.S. Moss (Elsevier, Amsterdam 1992); W.E. Pickett: Comput. Phys. Rep. **9**, 115 (1989)
7. R.M. Dreizler, E.K.U. Gross: *Density Functional Theory*. (Springer, Berlin Heidelberg New York 1990)

8. N. Troullier, J.L. Martins: Phys. Rev. B **43**, 1993 (1991)
9. S. Öğüt, J.R. Chelikowsky: Phys. Rev. Lett. **83**, 3852 (1999)
10. J.R. Chelikowsky, L. Kronik, I. Vasiliev, M. Jain, Y. Saad, in *Handbook of Numerical Analysis.* C. Le Bris (Ed.) Vol. 10: Computational Chemistry (Elsevier, Amsterdam 2003) pp. 613–637
11. M.E. Casida, in *Recent Advances in Density-Functional Methods.* Ed. by D.P. Chong (World Scientific, Singapure 1995) Pt. I, pp. 155–192
12. J.R. Chelikowsky, L. Kronik, I. Vasiliev: J. Phys. Condens. Matter **15**, 1517 (2003)
13. Takagahara, K. Takeda: Phys. Rev. B **53**, R4205 (1996)
14. A. Hill, S. Pokrant, A.J. Hill: J. Phys. Chem. B **103**, 3156 (1999)
15. F.A. Reboredo, A. Zunger: Phys. Rev. B **62**, R2275 (2000)
16. H.-C. Weissker, J. Furthmüller, F. Bechstedt: Phys. Rev. B **69**, 115310 (2004)
17. G. Nesher, L. Kronik, J.R. Chelikowsky: Phys. Rev. B **71**, 035344 (2005)
18. U. Itoh, Y. Toyoshima, H. Onuki, N. Washida, T. Ibuki: J. Chem. Phys. **85**, 4867 (1986)
19. B. Beschoten, P.A. Crowell, I. Malajovich, D.D. Awschalom, F. Matsukura, A. Shen, H. Ohno: Phys. Rev. Lett. **83**, 3073 (1999); Y. Ohno, D.K. Young, B. Beschoten, F. Matsukura, H. Ohno, D.D. Awschalom, Nature **402**, 790 (1999)
20. S.J. Pearton, C.R. Abernathy, D.P. Norton, A.F. Hebard, Y.D. Park, L.A. Boatner, J.D. Budai: Mater. Sci. Eng. R **40**, 127 (2003)
21. M. Jain, L. Kronik, J.R. Chelikowsky, V.V. Godlevsky: Phys. Rev. B **64**, 245205 (2001)
22. L. Kronik, M. Jain, J.R. Chelikowsky: Phys. Rev. B **66**, 041203 (2002)
23. M. Linnarsson, E. Janzen, B. Monemar, M. Kleverman, A. Thilderkvist: Phys. Rev. B **55**, 6938 (1997)
24. R.Y. Korotkov et al.: Appl. Phys. Lett. **80**, 1731 (2002)
25. E. Kojima, R. Shimano, Y. Hashimoto, S. Katsumoto, Y. Iye, M. Kuwata-Gonokami: Phys. Rev. B **68**, 193203 (2002)
26. F.E. Arkun, M.J. Reed, E.A. Berkman, N.A. El-Masry, J.M. Zavada, M.L. Reed, S.M. Bedair: Appl. Phys. Lett. **85**, 3809 (2004)
27. D. Loss, D.P. DiVincenzo: Phys. Rev. A **57**, 120 (1998)
28. M. Ouyang, D.D. Awschalom: Science **301**, 1074 (2003)
29. S. Sapra, D.D. Sarma, S. Sanvito, N.A. Hill: Nano Lett. **2**, 605 (2002)
30. S. Cho, S. Choi, S.C. Korea, Y. Kim, J.B. Ketterson, B. Kim, Y.C. Kim, J. Jung: Phys. Rev. B. **66**, 033303 (2002)
31. B. Oczkiewicz, A. Twardowski, M. Demianiuk: Solid State Commun. **64**, 107 (1987)
32. D.J. Norris, N. Yao, F.T. Charnock, T.A. Kennedy: Nano Letters **1**, 3 (2001)
33. X. Huang, J.R. Chelikowsky: Phys. Rev. B, in press
34. X. Huang, A. Makmal, J.R. Chelikowsky, L. Kronik: to be published
35. A. Stroppa, S. Picozzi, A. Continenza, A.J. Freeman: Phys. Rev. B **68**, 155203 (2003)
36. L.M. Sandratskii: Phys. Rev. B **68**, 224432 (2003)
37. These magnetic moments agree with a similar investigation of the three *bulk* alloys – T. Schulthess, W. Butler: J. Appl. Phys. **89**, 7021 (2001)
38. Y.D. Park, A.T. Hanbicki, S.C. Erwin, C.S. Hellberg, J.M. Sullivan, J.E. Mattson, T.F. Ambrose, A. Wilson, G. Spanos, B.T. Jonker: Science **295**, 651 (2002)
39. T. Dietl, H. Ohno, F. Matsukura, J. Cibert, D. Ferrand: Science **287**, 1019 (2000)
40. Y.-J. Zhao, T. Shishidou, A.J. Freeman: Phys. Rev. Lett. **90**, 047204 (2003)

8

Advances in Monte Carlo Simulations of Nanostructured Materials

G.C. Hadjisavvas and P.C. Kelires

Physics Department, University of Crete,
P.O. Box 2208, 710 03, Heraclion, Crete, Greece

Abstract. We discuss the application of novel Monte Carlo simulations to the study of nanostructured materials. Emphasis is given on interfacial properties and processes related to strain, disorder, and alloying. Two representative systems are investigated. We first apply a quasiequilibrium method to unravel the composition profiles in Ge islands formed on Si(100). We show that under near-equilibrium conditions the profile is largely dictated by the stress field in the interior of the islands and the tendency of Ge to segregate to the surface. The second problem discussed here is a composite system of Si nanocrystals embedded in a-SiO_2. By constructing realistic structural models and applying an efficient bond-switching algorithm, we obtain the equilibrium structure of the interface, and investigate its energetics, stability and disorder as a function of the nanocrystal size.

8.1 Introduction

As the size of the basic building blocks in nanostructured materials becomes smaller, the role of interfacial geometries in controlling the mechanical and optoelectronic properties attains crucial importance. At the interface, the bonding elements of the nanoparticles merge with those of the embedding medium. In this process, strain and disorder build up in the system, usually due to size mismatch and/or different merging environments. Thus, not only the topology, i.e., bond-length and bond-angle distortions, but also the chemical nature of the interface (bond types and their percentage) needs to be studied and if possibly controlled. Sharpness, disorder, and alloying at the interface are the main problems to be tackled. From the theoretical point of view, it is essential to employ the proper methods and techniques in order to unravel the equilibrium structure of the nanocomposite systems.

We review here recent work [1–4] aiming at an accurate description of interfacial atomistic processes in semiconducting nanomaterials, which is based on state-of-the-art Monte Carlo (MC) simulations. The employed techniques couple statistical precision with sufficiently accurate atomic interactions, enabling them to reach equilibrium in practical computational times, and to

Springer Proceedings in Physics, Volume 105
Computer Simulation Studies in Condensed-Matter Physics XVIII
Eds.: D.P. Landau, S.P. Lewis and H.-B. Schüttler

treat nanosystems of realistic dimensions. We discuss the application of these methods in two representative problems of high current interest: (i) The problem of intermixing in semiconductor quantum dots. The Ge/Si(100) case is taken as an example. (ii) The structure and energetics of Si nanocrystals embedded in amorphous silicon dioxide.

The transition from a planar, layer-by-layer two-dimensional (2D) growth mode to a 3D mode during semiconductor heteroepitaxy, characterized by the formation of nanometer-scale islands, usually referred to as quantum dots (QDs), is a result of the strain build up in the epilayer due to the size mismatch of the constituent elements [5]. In the Ge/Si(100) case, about three to four monolayers (ML) of Ge grow in a 2D manner on the Si substrate forming the wetting layer (WL), and subsequent deposited material forms nanoislands, pyramid- and/or dome-shaped, relaxing in a certain degree the epitaxial strain. While the island nucleation process [6,7], self-assembling and organization [8], and shape transitions [9] have been well studied, the issue of intermixing is not yet sufficiently understood.

Intermixing and its extent is expected to affect the quantum confinement within the QDs and their optoelectronic properties. It is therefore vital to gain knowledge about the composition variations within the islands. Not only mere mixing of species is an important factor, but also any variations of composition and the resulting inhomogeneities. Another related issue is the interlinking, if any, of interdiffusion with the stress field in the system. While there have been in the past numerous theoretical investigations of the stress field in QDs [7,10,11], no attempts to link stress and composition had been made.

One point of view is that interdiffusion lowers the effective lattice mismatch, and thus reduces the elastic strain energy (which partially remains even when islanding takes place). This is the *thermodynamic* model. Conclusions drawn depend on the assumptions made about the extent of atomic diffusion. Often, it is argued that only surface diffusion is significant, and so the composition is determined solely by the variation in strain across the growing island surface [12]. The equilibrium surface composition is buried by further growth, it becomes composition of the interior, and it freezes due to the lack of diffusion in the bulk of the island. Indications [13] that diffusion involves more layers near the surface, leads to extended equilibrium models, as the one presented here. On the other hand, the *kinetic* model assumes that the island composition is determined solely by kinematic factors and the random diffusion of adatoms on the surface [14].

The interpretation of experimental composition profiles [14–17] leads to rather controversial conclusions. It is still unclear whether these profiles, obtained under various deposition and annealing conditions, are a result of thermodynamics or kinetics, or both, and in what degree.

The second nanomaterial system discussed here, namely silicon nanocrystals (Si-nc) embedded in a-SiO_2, has been extensively studied in recent years because of its photoemission properties [18,19]. It is believed that the interface between the Si-nc and the oxide matrix plays a crucial role in controlling the optoelectronic properties. However, the interpretation of the origin of light emission relies on models which are mainly drawn from the consideration of isolated nanocrystals. These include models based on quantum confinement [20, 21], and on oxygen-related localized surface states [22–24].

Thus, despite its apparent importance, the structure of the interface of this composite material remains unclear. The kind and proportion of bonds, the width of the interface, and the Si oxidation states are crucial parameters which are poorly known. Also, it is essential to examine the stability of the nc in the amorphous oxide, against distortions and deformations, as they become smaller. This is particularly important for nc sizes below $\sim 2\,\mathrm{nm}$, where pinning of photoluminescence (PL) energies occurs [22–24].

We have recently reported [4] the first direct simulations of this system, which are based on realistic structural models and an efficient MC bond-switching algorithm. We obtained the equilibrium structure of the interface, and investigated its energetics and stability as a function of the nc size. Here, we discuss the most important findings of this study.

8.2 Methodology

For the study of Ge/Si QDs, we use a *quasi-equilibrium*, or we may say *constrained equilibrium* MC approach. We assume that at high enough temperatures, diffusion in the surface region, including the island, the WL, and few ML in the substrate is fast due to the strain. As a result, *local equilibrium* is established. This constrains us to sample over the local equilibria of the various metastable phases of the Ge/Si system in this region.

We work with a fixed system composition, i.e., we assume that a given amount of Ge atoms is deposited on Si(100) and this material forms the WL and the island. The formation of the composition profile is driven by free-energy minimization. Individual contributions include the surface energy, the strain energy, and the alloy mixing energy. Energy lowering is achieved by redistributing the atoms in the system. The chemical potential remains constant for fixed composition. Under these conditions, we equilibrate the system using the familiar isobaric-isothermal (N, P, T) ensemble, supplemented by Ising-type identity flips in the form of *mutual particle interchanges* (e.g. from Si to Ge at a certain site and vice versa at another site), so that the composition is kept constant [25]. We consider two possible types of switching moves. These are visualized in Fig. 8.1.

In the first type, the two atoms are chosen randomly. In the second type, which is the limiting case of the first type, we constrain the two atoms to be nearest neighbors. Obviously, equilibration is reached faster using the first type of moves, because diffusion of atoms at large distances is readily modeled. On the other hand, the second type is a more realistic representation of a diffusion event, but needs a considerably larger number of moves to diffuse an atom at large distances. By considering both types, we can test the consistency

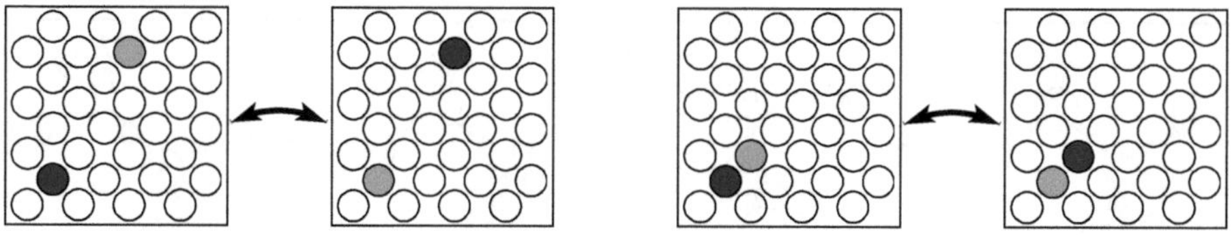

Fig. 8.1. Two different types of switching move: (**a**) two distant random atoms and (**b**) two nearest neighbors exchange identities

of the MC equilibration, since at the ergodic limit they ought to lead to similar island compositions. Note that we have no explicit barriers (activation energies) in the switching (exchange) moves of either type, besides the implicit barriers associated with overcoming the small size mismatch. We are currently working on the inclusion of such exchange barriers in the algorithm.

The implementation of this ensemble for MC simulations is done through the Metropolis algorithm. For details, see [26]. For the interactions, we use the well established interatomic potentials of Tersoff (T3) for multicomponent systems [27]. For some structures, we also use the Stillinger-Weber (SW) potential [28, 29], to make sure that the trends are independent of the energy functional.

The starting simulational cells consist of coherent pure Ge islands on top of the WL and the Si(100) substrate. We study both pyramid- and dome-shaped islands. The pyramids have a square base and {105} facets, aspect ratio (height over base width) $h/a \sim 0.1$, and contact angle $\sim 11°$. Their size is ~ 90 Å. They contain 1750 atoms. The domes are multifaceted, bounded by {113}, {105}, and {15 3 23} planes, and have an aspect ratio 0.2, and size 120 Å(7000 atoms.) The top layer of the WL and the dot facets are reconstructed in the 2×1 dimer configuration. The width of the WL is fixed at 3 ML [30]. The substrate contains 10 ML. The bottom layer is fixed, and identity switches occur down to 8 ML. Due to the limited depth of the Si substrate, the epitaxial strain is imposed by constraining laterally the cells to the Si lattice dimensions. Periodic boundary conditions are imposed in the lateral directions.

For the study of Si/a-SiO_2 nanocrystals, we used a MC methodology that has been well tested and applied with success for the description of the planar interface [31, 32]. In this method, the Si/a-SiO_2 system is modeled as a defect-free network in which Si and O have four and two bonds, respectively, without any O-O bonds. The energy is reasonably approximated by a Keating-like valence force model, given by

$$E_{\{\bar{r}\}} = \frac{1}{2}\sum_{i} k_b (b_i - b_0)^2 + \frac{1}{2}\sum_{i,j} k_\theta (\cos\theta_{ij} - \cos\theta_0)^2 + \gamma \sum_{m,n} (d_2 - |\bar{r_m} - \bar{r_n}|)^3 + U \,. \tag{8.1}$$

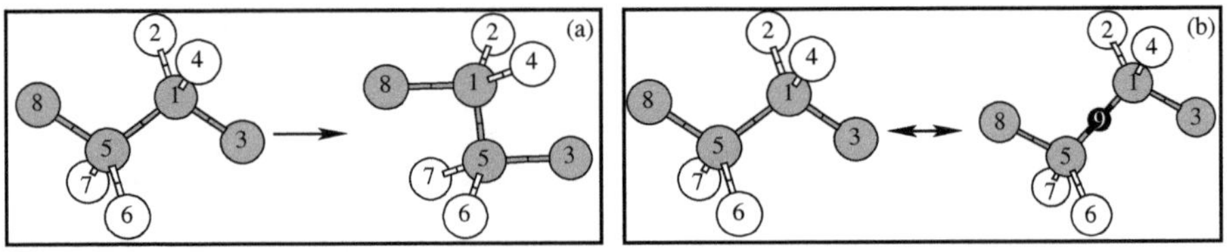

Fig. 8.2. Illustration of (**a**) WWW bond breaking and switching move and (**b**) bond conversion move

The first two terms represent the cost for bond-length and bond-angle distortions, respectively (strain energy). The third term enforces the restriction of two and four neighbors for O and Si, respectively. The term U is a "suboxide penalty", representing the chemical energy cost for the formation of any suboxide, taken from ab initio calculations [33]. Details about the parameters of the model can be found in [32].

To generate the amorphous oxide, the MC algorithm of Wooten, Winer, and Waire [34] is used. This is a well established method to generate continuous random networks, starting from the perfect crystal, by bond breaking and switching moves (Fig. 8.2a). To compositionally equilibrate the interface, we use bond conversion moves (Fig. 8.2b), which exchange a Si–Si bond in the nc with a neighboring Si-O-Si bond in the oxide. The number of Si and O atoms remains fixed. In both types of moves, we first relax locally the structure, following the attempted move, using a steepest-descent method minimizing the forces on the atoms. Then, the change ΔE in energy between this final and the initial configuration is calculated. The attempted move is accepted or rejected according to the Metropolis criterion. Bond switching and conversion moves are periodically followed by volume relaxation moves to relieve the stress in the entire system.

To generate the composite system, we start with a cubic cell in the β-crystobalite structure containing 8200 atoms. Within a radius from the center of the cell, all O atoms are removed, giving rise to an all-Si spherical region to simulate the Si-nc. Then, this starting geometry is relaxed to its energy minimum.

In the second stage, we perform MC bond switching in the oxide at a high temperature ($k_B T = 3\,\text{eV}$) for 40,000 moves, allowing it to liquify. Subsequently, the temperature is gradually reduced to 0.1 eV, bringing the oxide to its glassy, amorphous state. Quenching lasts for more than 1×10^6 moves. During these processes the positions of the atoms in the Si-nc are kept fixed. We then perform unconstrained MC bond conversion and switching of the entire composite system at 0.1 eV (887°C) for up to 3×10^6 moves, allowing it to both compositionally and topologically equilibrate.

We have generated in total 12 different fully relaxed composite structures. The size of the Si-nc ranges from 1.2 to 3.5 nm in diameter, and the number of atoms in the nc from ~ 30 to ~ 900. The total content of Si in the whole

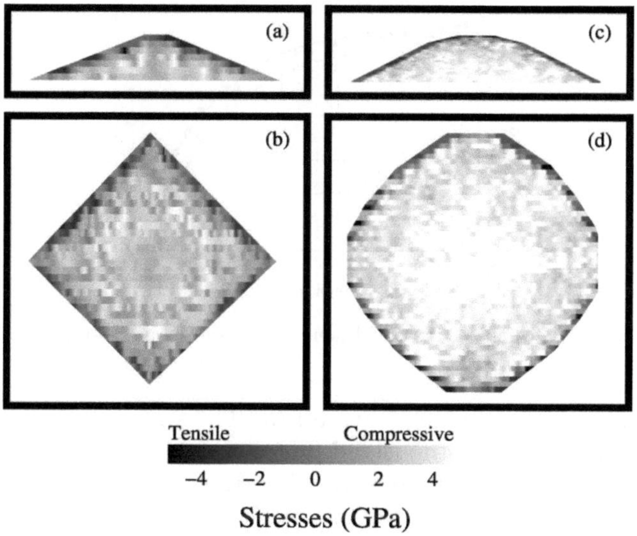

Fig. 8.3. Stress pattern of a pyramidal (*left*) and a dome island (*right*). Panels (**a**), (**c**) portray a thin slice cut through the island center. Panels (**b**), (**d**) show the base layer

structure varies from 34% to 44%, close to what is found experimentally. The properties of these structures are calculated by taking averages at 0.1 eV over 800,000 MC steps.

8.3 Results and Discussion

8.3.1 Ge/Si(100) Quantum Dots

We begin with the analysis of the stress field in pure Ge islands. This refers to situations where intermixing is negligible (relativily low growth temperatures), taking place during subsequent post-growth annealing. In our quasi-equilibrium model, the stress is one of the primary driving forces for intermixing (alloying). The stress pattern is probed using the concept of *atomic level stresses* [25]. The local stress can be viewed as an atomic hydrostatic compression (tension), defined by $\sigma_i = -dE_i/dlnV :\sim p\Omega_i$, where E_i is the energy of atom i (as obtained by decomposition of the total energy into atomic contributions), and V is the volume. Dividing by the appropriate atomic volume Ω_i converts into units of pressure p. Due to the definition of local stresses, positive (negative) sign indicates compressive (tensile) stress, respectively. Summing up the σ_i's over a group of atoms in the system (such as over a ML or over the whole island) yields the average stress over the specific configuration.

The stress pattern at 800 K, composed of atomic stresses, of two representative islands (a pyramid and a dome) is demonstrated in Fig. 8.3. These

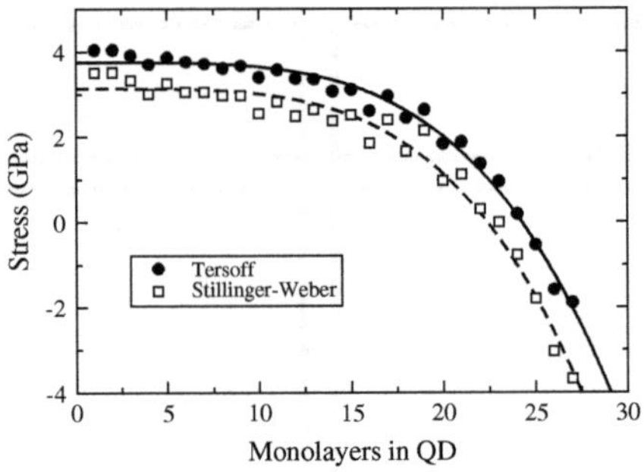

Fig. 8.4. Averager stress per layer in a dome island versus its height

graphs reveal three interesting features: (a) The stresses are overwhelmingly compressive in the core of the island, but compression fades as we move upwards to the top and laterally towards the edges. (b) Atoms near and around the periphery have neutral or slightly tensile stress. This is because stresses at the facets are tensile due to the dimer reconstruction. (c) The overall distribution of stress in the islands is inhomogeneous. Note, for example the ring of very compressed sites in the base of the pyramid [panel (a)]. These findings are independent of the potential used.

A layer-by-layer analysis of the stress in the dome island is given in Fig. 8.4. The symbols denote the average stress in a given layer. The lines portray the variation as we move from the base of the island (layer numbered 1) to the top. A non-linear decrease of compression takes place. Stress near the top varies (is released) more rapidly. Both potentials predict the same behavior, with the SW potential consistently suggesting somewhat lower stresses. (The average stresses over the whole island σ_{QD} are 3.4 and 2.7 GPa/atom for T3 and SW, respectively.) This shows that despite the three-dimensional formation and the accompanying outward atomic relaxations, the compressive stress in the dot is still substantial.

A similar analysis is carried for the WL. The average stresses are $\sim$ 2.5 GPa/atom, for both potentials, so despite the island formation on top the WL also remains substantially compressed. Most of the compression is stored in areas below the dot, especially around the periphery, forming a characteristic coral. This effect survives well into the substrate. The areas below the island in the substrate are under tension. Similar findings were as well reported by other researchers [11].

We model intermixing once the Ge islands have already been nucleated. Our approach is consistent with a wide class of post-growth annealing experiments reporting thermally-activated intermixing. This indicates a quasiequilibrium nature of the process. It also applies to the early stages of growth, when the stress built up in the island is not high enough to drive intermixing. There is experimental evidence [35] that small islands, such as those simulated here, alloy only after they have existed for some time after nucleating

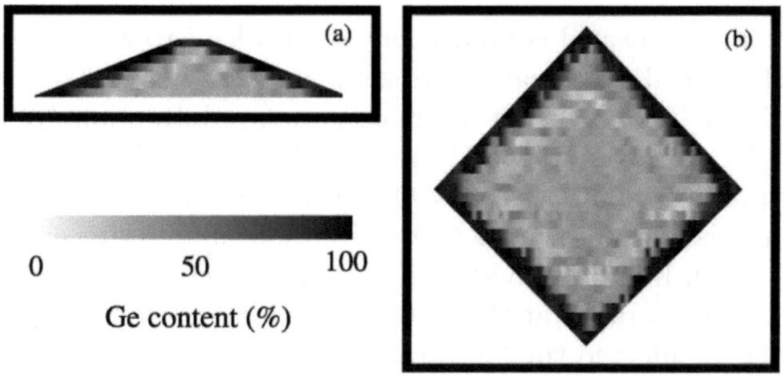

Fig. 8.5. Composition profile of a pyramidal island. (**a**) Thin slice cut through the island center. (**b**) Base layer

at medium growth temperatures (450–600°C). We also recently showed [3] that trenches (stress relieving grooves of missing material) around islands do not form until the middle stages of growth, and this precedes intermixing. Nevertheless, alloying during growth is an important issue, and it needs to be studied in detail (work is in progress).

The composition profiles in the islands are mapped by extracting the local compositions, which are indicated by the average site occupancies. These are calculated using the switcing-exchange moves depicted in Fig. 8.1. At the ergodic limit of many thousands attempted flips per site, both flip modes lead essentially to the same average occupancies. As an example, we show in Fig. 8.5 the profile of a pyramidal island calculated with the SW potential. Use of the T3 potential leads to similar results. Dome profiles have the same general characteristics and trends.

The central finding of this analysis is that the profile consists of a Si-rich core and an outer Ge-rich shell, which covers the dot from the base up to the top. This partition of the profile into two distinct regions is clear in both the vertical and lateral views. Yet, these regions are not completely homogeneous. In the base, for example, while the central region has heavily been enriched with Si, there are spots with higher Ge propabilities. By comparing to the stress patterns in Fig. 8.3, we readily see that the local stress conditions determine in a large degree the inhomogeneous features of the composition profile. This is because sites under compression (tension) tend to be occupied by the smaller (larger) species in the system [25]. The other important factor, that competes with the stress factor in shaping up the profile, is the lower surface energy of Ge. This factor is particularly strong in the periphery and at the faceted edges. By averaging the Ge fractions in every layer, we can extract the vertical variation of Ge content (not shown here). This comes out to have a slow variation through the island and only near the top varies rapidly. Lang et al. [36] have recently used a MC method similar to ours and extracted about the same profile.

Experimentally derived composition profiles lead to controversial conclusions. Malachias et al. [16] analyzed dome islands grown at 600°C by chemical vapor deposition (CVD), using anomalous x-ray diffraction (AXRD). They found a Si-rich core covered by a Ge-rich shell, and a rather slow vertical variation. Their profile is similar to ours. On the other hand, Schülli et al. [15] who analyzed dome islands grown at 600°C by molecular beam epitaxy (MBE), also using AXRD, found no evidence for a Si-rich core. The extracted variation of Ge content is rapid at the bottom, with an abrupt jump from $\sim 10\%$ to $\sim 80\%$. They neglected the lateral variations of composition in their analysis. Another interesting profile was suggested by Denker et al. [14], extracted mainly from pyramidal islands grown by MBE at 600°C using a selective etching technique. Instead of a Si-rich core, they proposed enriched Si-rich areas in the corners with the center remaining Ge rich. They explained this by a purely kinetic model of random surface diffusion of Si atoms, excluding strain-driven intermixing.

One may attempt to reconcile these differences by noting that the CVD method produces growth conditions near thermodynamic equilibrium in the surface region. This explains the similarity of the profile in [16] with our quasiequilibrium profile. If this is true, it implies that during CVD growth not only surface diffusion takes place but also some bulk diffusion below the central region of the island, in the form of Si-Ge exchanges, since this is required to explain the formation of a Si-rich core. In our method both types of exchanges are taken care of. The MBE method, instead, is a non-equilibrium process and it might mainly favor surface events, explaining the results in [14, 15]. Yet, it is puzzling that a recent study by Ratto et al. [17], who used MBE growth of Ge/Si(111), found the island centers rich in Si, proposing bulk driven alloying. Obviously, the issue of relative importance between surface and bulk diffusion mechanisms is open for further investigation.

8.3.2 Si/a-SiO_2 Nanocrystals

Let us now turn our attention to the Si/a-SiO_2 nc case. We first analyze the structure of the interface. Figure 8.6 shows a representative composite structure (a) before, and (b) after amorphization and full relaxation. A relatively large nc (3.2 nm) is illustrated. Our central finding is that a large number of Si-O-Si bridge bonds, in which an O atom connects two Si atoms terminating the Si-nc, have been formed at the interface. Their relative fraction is $\sim 60\%$. The driving force for their formation is the lowering of the interfacial strain energy. We estimate a drop in energy of $\sim 0.05\,\mathrm{eV/Å^2}$ when the interface relaxes by bond conversion, leading to enhancement of bridge bonding.

Bridge bonds have earlier been recognized [32,37] to lower the energy of the planar Si(001)/a-SiO_2 interface, because they can be stretched and bent with minimal energy cost [37]. Ordered arrays of bridges were proposed by ab initio calculations [37] and identified by MC simulations [32]. The level of bridge bonding reached in these simulations was of the order of 75%. It is remarkable

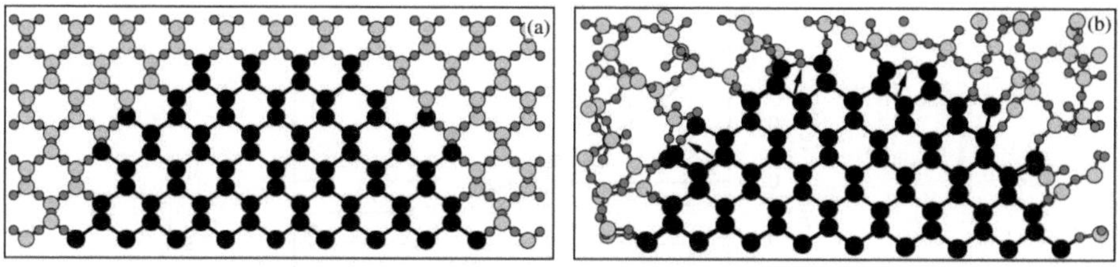

Fig. 8.6. Ball and stick model (part of a thin slice cut) of (**a**) the initial structure, and (**b**) of the final structure. *Dark spheres* show Si atoms in the nc. *Large (small) grey spheres* show Si (O) atoms in the oxide, respectively. Arrows indicate the formation of bridge bonds

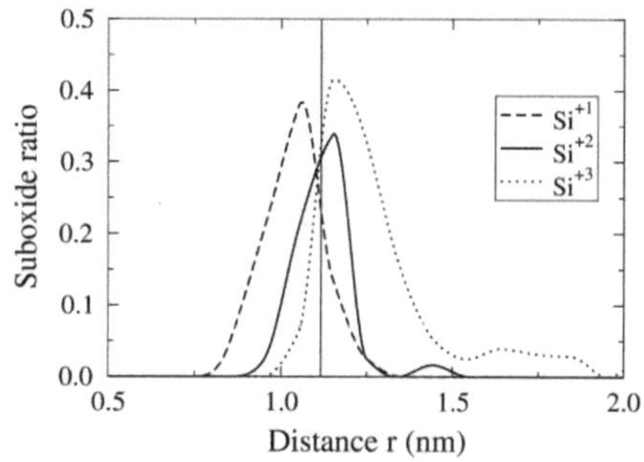

Fig. 8.7. The distribution of the suboxide components with respect to the center of the nanocrystal. The vertical solid line shows the nominal position of the interface

that the fraction of bridge bonds at the interface of Si-nc approaches that at the planar interface.

The distributions of suboxide components at the interface are shown in Fig. 8.7. We see that Si^{+2} is concentrated at the interface, Si^{+1} is found toward the Si-nc, and Si^{+3} occurs mainly toward the oxide, as in the case of the planar interface [32, 38]. The width of the transition layer containing the suboxides is ~ 0.8 nm, revealing a non-abrupt interface. This is consistent with the findings of the experimental work by Daldosso et al. [19].

The stability of the Si-nc as the size varies mainly depends on the interfacial energy. This energy can be defined as the difference between the total energy, including the suboxide penalty, and the sum of the bulk energies of the amorphous oxide and crystalline Si. Its variation is shown in Fig. 8.8. Overall, the energy rises as the Si-nc shrink, indicating lower stability. There are two distinctly different regimes in this variation. In the regime over 2–2.5 nm, the rise is slow, and the energies lie within the limits defined by the energy of the ideal planar Si(001)/a-SiO_2 interface with no suboxide layers (lower limit) and the energy of the planar interface with suboxide layers (upper limit) [32, 37].

On the other hand, the energies in the size regime below 2 nm rise sharply. Much of the sharp energy increase is due to strain which heavily builds up as

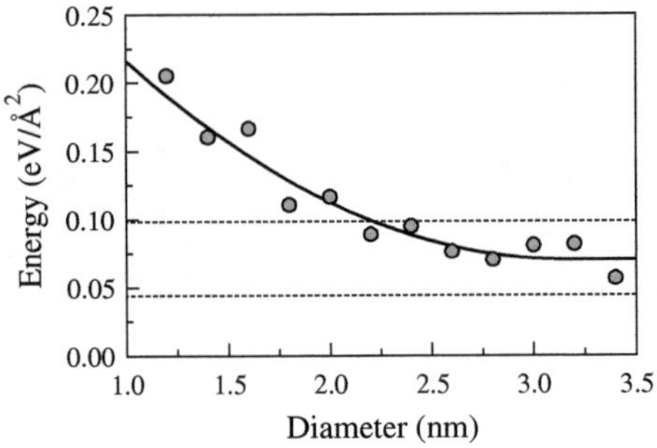

Fig. 8.8. Variation of interface energy with the Si-nc size. *Solid line* is a fit to the points. The *dashed horizontal lines* denote the energy of the planar interface without suboxides (*bottom*) and with suboxides (*top*)

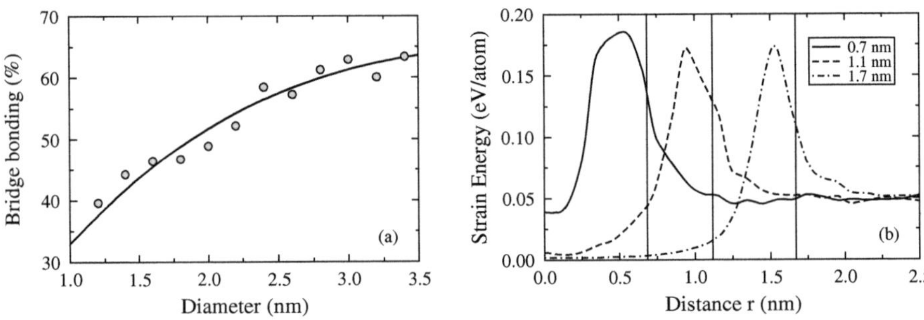

Fig. 8.9. (**a**) Percentage of bridge bonding as a function of Si-nc size. The *line* is a fit to the points. (**b**) The strain energy of three different nanocrystals versus the distance from their center. The *vertical solid lines* denote the nominal position of the respective interfaces

the Si-nc become smaller. There are two contributors to this increase of strain energy. The first is related to bridge bonding. As seen in Fig. 8.9a, the fraction of bridge bonds strongly declines as the size gets smaller, especially below 2–2.5 nm. Since bridge bonds are strain-relieving geometries, their reduction increases the interface energy.

The second contribution to the energy increase comes from the deformation incurred at the transition region, as a consequence of embedding the nc in the host oxide matrix. When these two different materials, having a large density gradient (9%), merge at their interface, the network topologies in either side deform in order to accomodate the transition. The smaller the nanocrystal, the more heavily strained and distorted is expected to be when put in the embedding host medium.

The effect of embedding on strain energy is exemplified in Fig. 8.9b, for three representative Si-nc. We observe a progressive increase of deformation and rise of strain energy as the size decreases. Most of the strain energy is concentrated inside the nanocrystals. The embedding oxide is less strained, because the amorphous network has more flexibility to accomodate the structural incompatibility. The striking feature is that the smaller nc (1.4 nm) is heavily deformed, even in the core region, which is reflected into its high interior strain energy.

The observation that bridge bonding declines and the nc deform as size decreases prompts us to give an alternative, although qualitative, explanation for the relatively small energy gaps in Si-nc in the size regime below 2 nm. It is possible that the observed PL redshift is not due to the interface oxygen states alone, but in a large degree also due to the high disorder in the nc. Strain and deformation in the nanocrystal introduce localized states near the band edges and effectively reduce the optical gap. A smaller contribution from the strained oxide side of the interface should also be anticipated [19]. A confirmation of this proposition requires electronic structure calculations.

Acknowledgement

This work is supported by a grant from the EU and the Ministry of National Education and Religious Affairs of Greece through the action "EΠE-AEK" (programme "HPAKΛEITOΣ".)

References

1. P. Sonnet, P.C. Kelires: Phys. Rev. B **66**, 205307 (2002)
2. G.C. Hadjisavvas, P. Sonnet, P.C. Kelires: Phys. Rev. B **67**, 241302(R) (2003)
3. P. Sonnet, P.C. Kelires: Appl. Phys. Lett. **85**, 203 (2004)
4. G.C. Hadjisavvas, P.C. Kelires: Phys. Rev. Lett. **93**, 226104 (2004)
5. J. Stangl, V. Holy, G. Bauer: Rev. Mod. Phys. **76**, 725 (2004)
6. J. Tersoff, R.M. Tromp: Phys. Rev. Lett. **70**, 2782 (1993); J. Tersoff and F.K. LeGoues: *ibid.* **72**, 3570 (1994)
7. A. Barabási: Appl. Phys. Lett. **70**, 2565 (1997)
8. J. Tersoff, C. Teichert, M.G. Lagally: Phys. Rev. Lett. **76**, 1675 (1996)
9. F.M. Ross, J. Tersoff, R.M. Tromp: Phys. Rev. Lett. **80**, 984 (1998)
10. W. Yu, A. Madhukar: Phys. Rev. Lett. **79**, 905 (1997)
11. P. Raiteri, L. Miglio, F. Valentinotti, M. Celino: Appl. Phys. Lett. **80**, 3736 (2002)
12. N. Liu, J. Tersoff, O. Baklenov, A.L. Holmes, C.K. Shih: Phys. Rev. Lett. **84**, 334 (2000)
13. K. Nakajima, A. Konishi, K. Kimura: Phys. Rev. Lett. **83**, 1802 (1999)
14. U. Denker, M. Stoffel, O.G. Schmidt: Phys. Rev. Lett. **90**, 196102 (2003)
15. T.U. Schülli, J. Stangl, Z. Zhong, R.T. Lechner, M. Sztucki, T.H. Metzger, G. Bauer: Phys. Rev. Lett. **90**, 066105 (2003)
16. A. Malachias, S. Kycia, G. Medeiros-Ribeiro, R. Magalhaes-Paniago, T.I. Kamins, R.S. Williams: Phys. Rev. Lett. **91**, 176101 (2003)
17. F. Ratto, F. Rosei, A. Locatelli, S. Cherifi, S. Fontana, S. Heun, P. Szkutnik, A. Sgarlata, M. De Crescenzi, N. Motta: Appl. Phys. Lett. **84**, 4526 (2004)

18. L. Pavesi, L. Dal Negro, C. Mazzoleni, G. Franzo, F. Priolo: Nature (London) **408**, 440 (2000)
19. N. Daldosso *et al.*: Phys. Rev. B **68**, 085327 (2003)
20. J.P. Proot, C. Delerue, G. Allan: Appl. Phys. Lett. **61**, 1948 (1992)
21. L.W. Wang, A. Zunger: J. Phys. Chem. **98**, 2158 (1994)
22. M.V. Wolkin, J. Jorne, P.M. Fauchet, G. Allan, C. Delerue: Phys. Rev. Lett. **82**, 197 (1999)
23. A. Puzder, A.J. Williamson, J.C. Grossman, G. Galli: Phys. Rev. Lett. **88**, 097401 (2002)
24. I. Vasiliev, J.R. Chelikowsky, R. Martin: Phys. Rev. B **65**, 121302(R) (2002)
25. P.C. Kelires, J. Tersoff: Phys. Rev. Lett. **63**, 1164 (1989); P.C. Kelires: *ibid.* **75**, 1114 (1995)
26. P.C. Kelires: J. Phys. Condens. Matter **16**, S1485 (2004)
27. J. Tersoff: Phys. Rev. B **39**, 5566 (1989)
28. F. Stillinger, T. Weber: Phys. Rev. B **31**, 5262 (1985)
29. P.C. Kelires: Phys. Rev. B **49**, 11496(R) (1994)
30. Y.W. Mo, D.E. Savage, B.S. Swartzentruber, M.G. Lagally: Phys. Rev. Lett. **65**, 1020 (1990)
31. K.O. Ng, D. Vanderbilt: Phys. Rev. B **59**, 10132 (1999)
32. Y. Tu, J. Tersoff: Phys. Rev. Lett. **84**, 4393 (2000); *ibid.* **89**, 086102 (2002)
33. D.R. Hamann: Phys. Rev. B **61**, 9899 (2000)
34. F. Wooten, K. Winer, D. Weaire: Phys. Rev. Lett. **54**, 1392 (1985)
35. S.A. Chaparro, J. Drucker, Y. Zhang, D. Chandrasekhar, M.R. McCartney, D.J. Smith: Phys. Rev. Lett. **83**, 1199 (1999)
36. C. Lang, D. Nguyen-Manh, D.J.H. Cockayne: J. Appl. Phys. **94**, 7067 (2003)
37. R. Buczko, S.J. Pennycook, S.T. Pantelides: Phys. Rev. Lett. **84**, 943 (2000)
38. J.H. Oh *et al.*: Phys. Rev. B **63**, 205310 (2001)

9

Molecular Dynamics Study of Atomic Displacements and X-Ray Diffuse Scattering

Y. Puzyrev and J.S. Faulkner

Alloy Research Center, Department of Physics, Florida Atlantic University, Boca Raton FL 33431, USA

Abstract. Molecular Dynamics calculations of atomic displacements in pure copper and copper-gold alloys were performed to study atomic size effect. The alloys was chosen due to a large size mismatch and the existence of embedded-atom potentials. The potentials are corrected to match the lattice constants. The nearest neighbor bond length crossover is calculated in all systems. The theoretical diffuse scattering intensity agrees well with the experimentally observed values for pure copper.

9.1 Theory

For the binary alloys with components close to each other in periodic table, for example, iron-nickel alloy [1], the thermal diffuse scattering can be removed. It can be done by subtracting the intensity of the scattering for X-ray beam with energy that makes the scattering factors of iron and nickel equal. This energy does not exist for copper-gold alloy, since copper and gold have very different scattering factors. Therefore, the recovery of chemical short-range order parameters and atomic displacements from the diffuse scattering intensity for this alloy and alloys with large size mismatch is complicated by the intensity redistribution due to the thermal vibrations of atoms. The explanation of the properties of solids based of density functional theory local density approximation (DFT-LDA) [2] calculations of the electronic states has been so successfull that, on the occasion that there is not complete agreement with experimental results, scientists look for something that was left out of the calculation. Some of the groups that are involved in developing first-principle theories for metallic alloys are now making an effort to incorporate atomic displacements (AD) into their methods [3].

9.1.1 Atomic Displacements

The AD in small clusters of atoms have been calculated with DFT-LDA cal-cualtions [4,5] and lead to the conflicting results. The embedded-atom poten-

Springer Proceedings in Physics, Volume 105
Computer Simulation Studies in Condensed-Matter Physics XVIII
Eds.: D.P. Landau, S.P. Lewis and H.-B. Schüttler

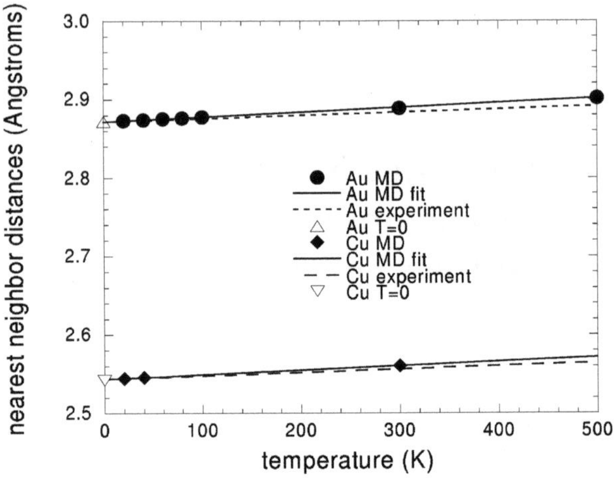

Fig. 9.1. The nearest neighbor distances in pure Cu calculated at the indicated temperatures with MD are shown by the *diamonds*, and those for Au by the *filled circles*. *Straight line* fits to these data are shown by the *solid lines*. The *triangles* at $T = 0$ are calculated with the CG method. The *dotted lines* are fits to the experimental data

tials [6] are used for the calculations on large supercells [7]. These calculations use a model where atoms have no motion other then static displacements. The AD obtained in these calculations are compared to the most frequently quoted EXAFS paper [8] which purports to measure the AD in *concentrated* metallic alloys. It would be extremely useful to have reliable evidence about the AD in metallic alloys, and the most accurate way to do this experimentally is to analyze the diffuse scattering data taken with X-rays at a synchrotron source such as the APS in Argonne, Illinois [9].

9.1.2 Atomic Size Effects and Diffuse Scattering

Atomic size effects have been studied by X-ray and neutron diffractionists for decades. The thermal motion of atoms in pure alloys lead to the reduction in the intensity of the Bragg peak located at the scattering angle θ. This reduction is exponential with the distance from the origin and is given by the factor

$$\exp\left[\frac{-2(B+B')\sin^2\theta}{\lambda^2}\right],$$

where B is the standard Debye–Waller factor and B' is a static Debye–Waller factor. These factors are proportional to mean square displacements of the atoms about their ideal sites due to the temperaturedependent excitation of phonons $\langle u_{th}^2\rangle$, and the mean square of the static atomic displacements

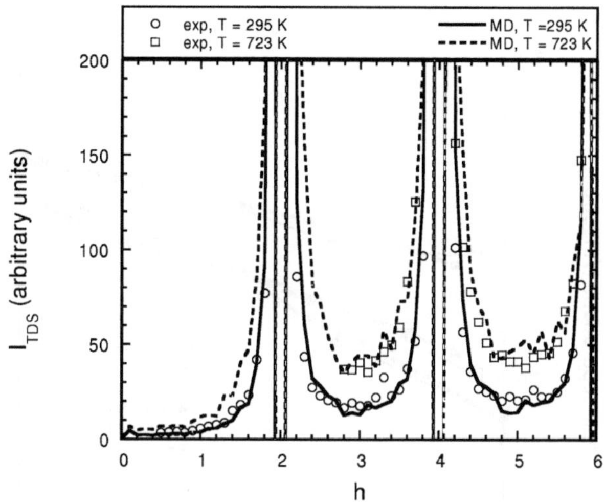

Fig. 9.2. A graph of the temperature diffuse scattering for pure copper at temperatures 295 K and 723 K along the direction (h00) in k-space. The experimental intensities, shown by *circles* and *squares*, are from the unpublished communication quoted in [13]. The intensities calculated with MD are shown by the *solid* and *dotted lines*. The units of the intensities are arbitrary, and h is in units of 2π over the lattice constant

$\langle u_{st}^2 \rangle$. For the case of an alloy with very large size mismatch, copper-gold, it was found experimentally [10] that B' is equal to B at liquid nitrogen temperatures and $B' = B/3$ at room temperature. If copper and gold retained the sizes that they have in pure metals, B' would be twenty times as large as one observed. The data analysis in this experiment has an assumption that elements have the same thermal displacements. This assumption is not obvious and, therefore, we are calculating the displacements of atoms in an alloy at room temperature and above using molecular dynamics. From this we will obtain nearest neighbor distances as a function of concentration. The arrangements of atoms that arise in the calculation will be used to calculate the diffuse scattering maps directly. We are using EAM potentials [6] and ParaDyn molecular Dynamics code [11].

9.2 Results of Calculations

Calculations were performed on 32,000 atoms using a NPT ensemble with periodic boundary conditions for force calculations. Hoover damping [12] was chosen as a temperature control mode. For each temperature and concentration number of MD steps must be allowed for thermal expansion until equilibrium is reached.

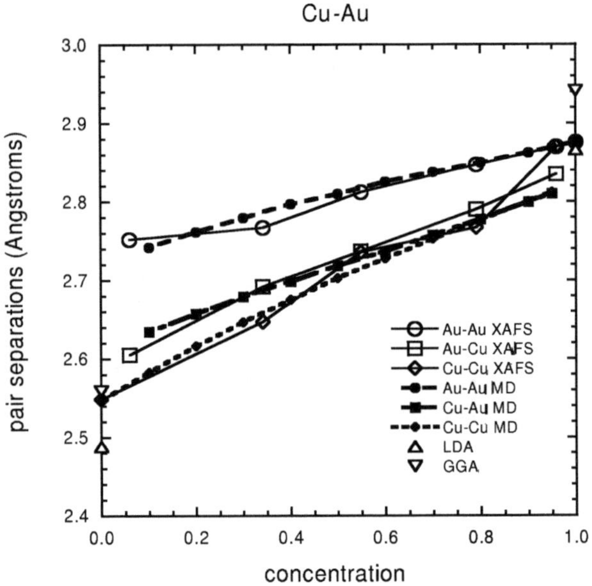

Fig. 9.3. The Cu-Cu, Cu-Au, and Au-Au pair separations as a function of concentration for Cu-Au alloys are shown with *filled diamonds*, *squares*, and *circles*. The data measured with XAFS are indicated by the *open markers*. The *open triangles* show the results of LDA and GGA calculations

Thermal Expansion

From Fig. 9.1 we can see the difference in atomic sizes of copper and gold as well as the thermal expansion of these two elements. Neither LDA nor CGA approximations in DFT calculations give the lattice constants of both copper and gold with desired accuracy. It is clear that MD predicts thermal expansions well, and therefore it can be used for the evaluation of thermal and static isplacements. The dependence of diffuse scattering intensity on the position in reciprocal space is in a good agreement with experiment as shown in Fig. 9.2. This indicates that the EAM potentials describe the atomic size and the atomic displacements adequately. Note that the calculations are performed at the same temperature as the experiments [13], i.e. room temperature.

Nearest Neighbor Distances

As shown in Fig. 9.3 the nearest neighbor distances are in excellent agreement with the experimental results. The copper-gold distance becomes equal to the copper-copper distance at 86% gold, which is observed in XAFS experiment. The existence of the crossover indicates that the many-body interaction plays important part in the description of metallic alloys.

9.2.1 Conclusions

- Simple classical Molecular Dynamics with embedded-atom potentials predicts well thermal expansion in alloys in contrast to the first principles calculations.
- The diffuse scattering can be reproduced from the atomic positions obtained in Molecular Dynamics calculations.
- The position of copper-copper and copper-gold bond lengths crossover in alloy is in a good agreement with experiment.

In the future we plan to

- Use the instanteneous atomic positions from MD calculation to predict the diffuse scattering maps for alloys measured experimentally.
- Treat other alloys both experimentally and theoretically using MD.

References

1. G.E. Ice, C.J. Sparks, A. Habenschuss, L.B. Shaffer, Phys. Rev. Lett. **68**, 863 (1992)
2. P. Hohenberg, W. Kohn, Phys. Rev. **136**, B864 (1964); W. Kohn, L.J. Sham, Phys. Rev. **140**, A1133 (1965)
3. Z.W. Lu, S.H. Wei, A. Zunger, Phys. Rev. B **44**, 3387 (1991)
4. L.G. Fereira, V. Ozolins, A. Zunger, Phys. Rev. B **60**, 1687 (1998)
5. A.V. Ruban, S.I. Simak, S. Shallcross, H.L. Skriver, Phys. Rev. B **67** 214302 (2002)
6. M.S. Daw, S. Foiles, M.I. Baskes, Mater. Sci. Rep. **9**, 251 (1993)
7. N. Mousseau, M.F. Thorpe, Phys. Rev. B **45**, 2015 (1992)
8. A.I. Frenkel, E.A. Stern, A. Rubstein, A. Voronel, Yu. Rosenberg, J. Phys. IV France **7**, C2-1005 (1997)
9. G.E. Ice *et al.*, J. Phase Equilibria **19**, 529 (1998)
10. F.H. Herbstein, B.S. Borie, Jr., B.L. Averbach, Acta Cryst. **9**, 466 (1956)
11. S.J. Plimpton, B.A. Hendrickson, in *Materials Theory and Modelling*, ed. by J. Broughton, P. Bristowe, J. Newsam. (MRS Proceedings 291, Pittsburgh, PA 1993) p. 37
12. W.G. Hoover, Phys. Rev. A **31**, 1695 (1985)
13. C.J. Sparks, G.E. Ice, private communication

Part III

Phase Transition and Other Topics in Statistical Mechanics

10

A Note on Stable States of Dipolar Systems at Low Temperatures

Y. Tomita[1], K. Matsushita[1], A. Kuroda[1], R. Sugano[2], and H. Takayama[1]

[1] Institute for Solid State Physics, University of Tokyo, Kashiwa, Chiba 277-8581, Japan
[2] Advanced Research Laboratory, Hitachi, Ltd., Hatoyama-machi, Saitama 350-0395, Japan

10.1 Introduction

In the past several years, many important innovations in nanotechnology were made. Today it becomes possible to make nanosize magnetic particles, and development of high storage-density magnetic device is desired. Though dipole interaction plays the main role in these magnetic particle systems, there is little systematic study of dipolar systems [1].

Luttinger and Tisza (LT) discussed the ground states of three-dimensional dipole cubic lattices in their pioneering work [2]. They assumed $\boldsymbol{\Gamma}^2$ symmetry that ground states are constructed by the translations of the dipoles on $2\times2\times2$ cubic unit, and examined its dipole configurations. In addition, since the dipole interaction energy is a quadratic form, even though dipole moment has $O(3)$ symmetry, it is sufficient to consider only eight arrays for each component of the moment. We define eight arrays by

$$\boldsymbol{A}_{b_x b_y b_z}(\boldsymbol{l}) = (-1)^{b_x l_x + b_y l_y + b_z l_z}\,, \quad (b_x, b_y, b_z, l_x, l_y, l_z = 0, 1)\,. \tag{10.1}$$

Here b_x, b_y, b_z are indices for a dipole configuration, and an array (l_x, l_y, l_z) represents a corner of the unit cube. An eight-dimensional vector $\boldsymbol{A}$ corresponds to $\boldsymbol{X}$, $\boldsymbol{Y}$, or $\boldsymbol{Z}$ depending on its component of moments. In Fig. 10.1, we show the eight basic arrays.

For the simple cubic (SC) lattice, Luttinger and Tisza predicted the columnar antiferromagnetic state,

$$a\boldsymbol{X}_{011} + b\boldsymbol{Y}_{101} + c\boldsymbol{Z}_{110}\,, \quad (a^2 + b^2 + c^2 = 1)\,, \tag{10.2}$$

has the lowest energy. For the body centered cubic (BCC) lattice, lowest energy states predicted by LT are tabulated in Table 10.1.

Springer Proceedings in Physics, Volume 105
Computer Simulation Studies in Condensed-Matter Physics XVIII
Eds.: D.P. Landau, S.P. Lewis and H.-B. Schüttler

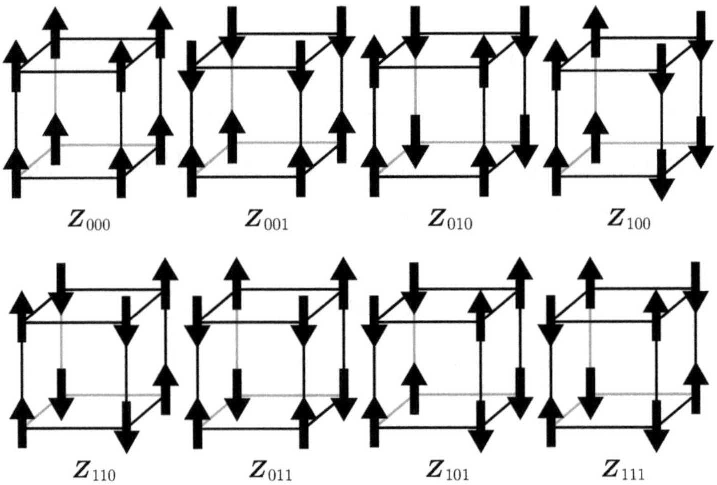

Fig. 10.1. The eight basic arrays

Table 10.1. Lowest energy states are tabulated below. The states of lattice points (l.p.) and body centers (b.c.) are commutative. The condition, $a^2 + b^2 + c^2 = 1$, is satisfied for the first line, and others satisfy the condition, $a^2 + b^2 = 1$

At l.p. (b.c.)	At b.c. (l.p.)
$a\boldsymbol{X}_{101} + b\boldsymbol{Y}_{110} + c\boldsymbol{Z}_{011}$	$b\boldsymbol{X}_{110} + c\boldsymbol{Y}_{011} + a\boldsymbol{Z}_{101}$
$a\boldsymbol{X}_{101} + b\boldsymbol{Z}_{101}$	$b\boldsymbol{X}_{101} + a\boldsymbol{Z}_{101}$
$a\boldsymbol{Y}_{110} + b\boldsymbol{X}_{110}$	$b\boldsymbol{Y}_{110} + a\boldsymbol{X}_{110}$
$a\boldsymbol{Z}_{011} + b\boldsymbol{Y}_{011}$	$b\boldsymbol{Z}_{011} + a\boldsymbol{Y}_{011}$

10.2 Results

In order to examine the lowest energy states predicted by LT, we simulate dipolar systems on the SC lattice and the BCC lattice at low temperatures. The Hamiltonian for system size L is given by

$$\mathcal{H} = \sum_{n_x,n_y,n_z} \sum_{i<j} \left[\frac{\boldsymbol{\mu}_i \cdot \boldsymbol{\mu}_j}{(\boldsymbol{r}_{ij} + \boldsymbol{n}L)^3} - 3\frac{(\boldsymbol{\mu}_i \cdot (\boldsymbol{r}_{ij} + \boldsymbol{n}L))(\boldsymbol{\mu}_j \cdot (\boldsymbol{r}_{ij} + \boldsymbol{n}L))}{(\boldsymbol{r}_{ij} + \boldsymbol{n}L)^5} \right] . \tag{10.3}$$

We employed heat bath method for Monte Carlo spin update, and Ewald summation method [3–5] is used for counting long range dipolar interaction. As expected, for the SC and BCC of $L = 2$ lattice systems, dipole configurations produced by simulations well agree with the prediction by LT. The dipole configurations are consistent with the lowest energy state described in (10.2) for the SC of $L \geq 4$ lattices. On the other hand, for the BCC of $L \geq 4$ lattices, the stable dipole configuration that we obtained are different from the states in Table 10.1, which is shown Fig. 10.2b.

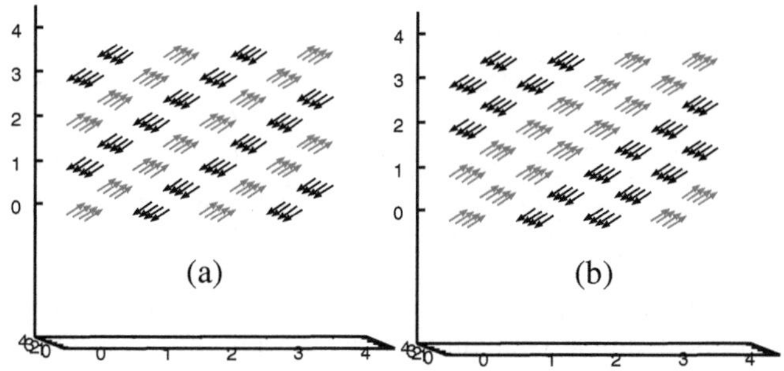

Fig. 10.2. Stable states for the body centered cubic lattice. (**a**) A stable state which is consistent with LT's predicted ground states. (**b**) A stable state obtained by MC simulation which is not consistent with the states in Table 10.1

Table 10.2. The zero temperature internal energy dependence on thickness of ferromagnetic layer

Thickness of FM layer	$L = 2$	$L = 4$	$L = 6$	$L = 8$
1	−7.94368	−7.94368	−7.94368	−7.94368
2	–	−8.16968	–	−8.16968
3	–	–	−8.23896	–
4	–	–	–	−8.27087

The dipole configuration in Fig. 10.2b is characterized by stacked dipole layers in which all dipoles align in the same direction. Such stacked dipole layer configurations can be constructed from LT's predicted state, although ferromagnetic layer is single (Fig. 10.2a), but not double as in Fig. 10.2b. We calculate the zero temperature internal energy as a function of thickness of ferromagnetic layer to check our result of MC simulation. The results are shown in Table 10.2. At least till $L = 8$, the internal energy becomes lower as the thickness of ferromagnetic layer increases. Furthermore the values of these internal energies are lower than that of LT calculated. This means that one cannot assume $\boldsymbol{\Gamma}^2$ symmetry to dipole systems on the BCC lattice.

We set up a simplified model to obtain an intuitive understanding of stable states of dipolar systems at low temperatures. In some cases, consideration of low dimensional system provides insights into physics of the system. So we start looking at two-dimensional systems.

Typical ground states of the square lattice and triangular systems are depicted in Fig. 10.3. In order to investigate the stable state at low temperatures, we simplify the model as follows: we treat the system as interacting ferromagnetic dipole chain system, and ferromagnetic chain can take only two values (rightward or leftward as in Fig. 10.3). Then, Hamiltonian becomes

$$\mathcal{H} = \sum_{i \neq j} f(r_{ij}) \sigma_i \sigma_j \, , \quad (\sigma = \pm 1) \, , \tag{10.4}$$

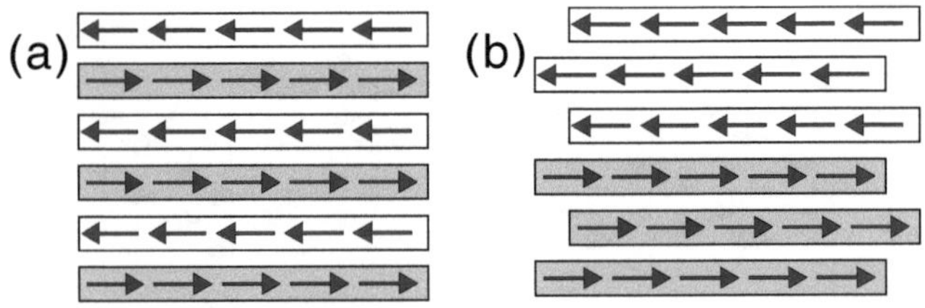

Fig. 10.3. (**a**) A typical dipole configuration in the ground state on the square lattice. (**b**) A typical dipole configuration in the ground state on the triangular lattice

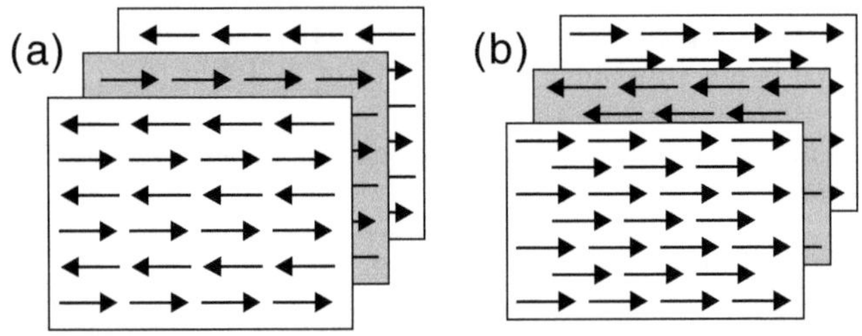

Fig. 10.4. (**a**) A typical dipole configuration in the ground state on the SC lattice. (**b**) A typical dipole configuration in the ground state on the BCC lattice

that is, the model is mapped to a one-dimensional Ising chain with long-range interaction $f(r)$. We estimate $f(r)$ by numerical calculation with $L/2$ cut off. At the zero temperature internal energy of the square lattice and the triangular lattice was calculated for several ferromagnetic dipole chain configurations. For the square lattice, antiferromagnetic state is the most stable, whereas for the triangular lattice, antiparallel two ferromagnetic domains structure is the most stable. This difference comes from the shape of $f(r)$, that is, $f(r)$ for square lattice is positive and monotonically decreasing function. For the triangular lattice, on the other hand, except for nearest neighbor, $f(r)$ is positive and its absolute value decreases monotonically. That is to say, ordering in the same direction gains energy in short range, while ordering in opposite direction gains energy in long range; as a consequence, ferromagnetic domains structure is realized at low temperatures.

Since the simplified model succeeded in explaining dipole configurations on two-dimensional lattices at low temperatures, we extend the model to three-dimensional systems. In Fig. 10.4, we show typical ground states for the SC lattice and the BCC lattice.

If one notes dipole configuration on the gray sheet is merely reversed configuration on the white sheet, it is natural to apply the same tactics which is employed in the two-dimensional system to the three-dimensional systems: there are magnetic sheets interacting each other, and a magnetic sheet can take only two values as like Ising variable. Then, we obtain simplified Hamiltonian, $\mathcal{H} = \sum_{\rangle \neq |} \{(\nabla_{\rangle|})\sigma_{\rangle}\sigma_{|}$. Again, the model is mapped to a one-dimensional

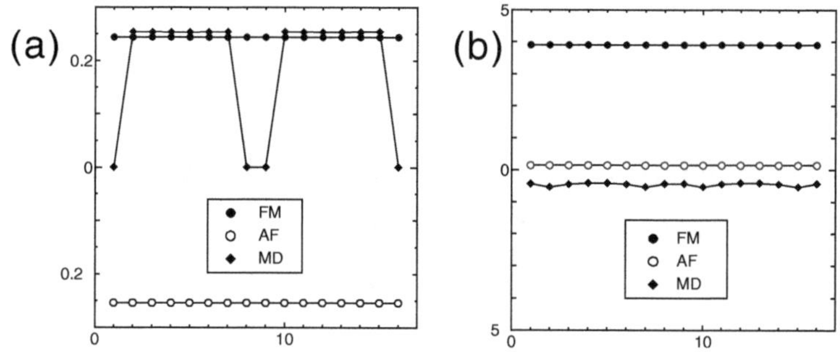

Fig. 10.5. (**a**) The internal energy at each site (layer) of the SC lattice ($L = 16$). The internal energy per site is $E_{\mathrm{FM}} = 0.2437$, $E_{\mathrm{AF}} = -0.2437$, and $E_{\mathrm{MD}} = 0.1907$ for ferromagnetic dipole configuration, antiferromagnetic dipole configuration, and magnetic domain dipole configuration respectively. (**b**) The internal energy at each site (layer) of the BCC lattice ($L = 16$). The internal energy per site is $E_{\mathrm{FM}} = 3.9016$, $E_{\mathrm{AF}} = 0.1488$, and $E_{\mathrm{MD}} = -0.4508$ for each dipole configuration

Ising chain with long-range interaction. In Fig. 10.5, we show the results of the SC lattice and the BCC lattice.

The result of the SC lattice supports the one of LT. On the other hand, the result of the BCC lattice is not consistent with the LT result, but with our results of MC simulation. It is notable that ferromagnetic domains structure is stable for the BCC lattice, even though interaction $f(r)$ is positive. We also examined $L = 64$ chain model, and we confirmed the four ferromagnetic domains structure is the most stable.

10.3 Summary

To summarize, we showed "antiferromagnetic structure" is stable for the square lattice and the SC lattice as LT predicted. On the other hand, for the triangular lattice and the BCC lattice, magnetic domain structure is stable. Theoretical approach which assumes $\boldsymbol{\Gamma}^2$ symmetry fails in the triangular lattice and the BCC lattice. For future works, examination of the face centered cubic lattice and estimation of critical temperatures and exponents for several lattices are remained.

References

1. S. Romano: Nuovo Cimento D **7**, 717 (1986)
2. J.M. Luttinger, L. Tisza: Phys. Rev. **70**, 954 (1946)
3. P.P. Ewald: Ann. Phys. (Paris) **64**, 253 (1921)
4. H. Kornfeld: Z. Phys. **22**, 27 (1924)
5. S.W. de Leeuv, J.W. Perram, E.R. Smith: Proc. R. Soc. London Ser. A **373**, 27, 57 (1980)

11

Low-Temperature Phase Boundary of Dilute Lattice Spin Glasses

S. Boettcher

Physics Department, Emory University, Atlanta, Georgia 30322, USA

Abstract. Based on a recently developed algorithm for the exploration of ground states in bond diluted lattice spin glasses, we determine the scaling of defect energies with system size for Ising spin glasses at the bond percolation threshold p_c. The results can be related by well-known scaling relations to the shape of the transition temperature $T_g \sim (p - p_c)^\phi$ between the paramagnetic and glassy regime for $p \to p_c$. The numerical results for in three dimensions are consistent with rather old experimental data for $(Fe_xNi_{1-x})_{75}P_{16}B_6A_3$, suggesting that new experimental work may be able to put those numerical predictions to the test.

The exploration of low-temperature properties of disordered systems remains an important and challenging problem [1]. Systems in this class possess a low-temperature glassy state with a glass transition at some temperature $T_g > 0$. They are characterized by a complex (free-)energy landscape in configuration space with a hierarchy of valleys and barriers whose multimodal structure impedes the progression of any dynamics towards equilibration, causing tantalizing phenomena, such as trapping and jamming on intermediate time scales, and aging on long time scales. An understanding of such systems is of paramount importance as these phenomena are observed for a large class of materials as well as for biological systems [2].

The paradigmatic model for the study of such phenomena is the Ising spin glass on a lattice (Edwards-Anderson model, EA [3]),

$$H = -\sum_{<i,j>} J_{i,j}\, x_i\, x_j\,, \quad (x_i = \pm 1)\,. \tag{11.1}$$

Disorder effects arise via quenched random bonds, $J_{i,j}$, mixing ferromagnet and anti-ferromagnetic couplings between nearest-neighbor spins, that lead to conflicting constraints which leave variables frustrated. It is believed that a proper understanding of static and dynamic features of EA may aid a description of the unifying principles expressed in a class of materials [4]. Most insights into finite-dimensional systems has been gained through computational approaches that elucidate low-temperature properties [2, 5–7].

Springer Proceedings in Physics, Volume 105
Computer Simulation Studies in Condensed-Matter Physics XVIII
Eds.: D.P. Landau, S.P. Lewis and H.-B. Schüttler

Here, we will extract the response induced through defect-interfaces [8] near $T = 0$. These can be created by fixing the spins along the two faces of an open boundary in one lattice direction. The ground state configuration with energy E_0 of an instance is first determined for a random fixing of those boundary spins, then the energy E_0' is obtained for the same instance and the same fixing, but with all spins reversed on one of the faces. Hence, the interface energy $\Delta E = E_0' - E_0$ created by the perturbation on the boundary is sampled, and its distribution $P(\Delta E)$ determined. In most inquiries of this type, one is interested to see if a system possesses a glassy ordered state for low T. Then, the typical energy scale involved, here represented by the width of the distribution, $\sigma(\Delta E) = \sqrt{\langle \Delta E^2 \rangle - \langle \Delta E \rangle^2}$, should grow with the size of the perturbation, say, the linear extend of the boundary, L, a $\sigma(\Delta E) \sim L^y$. This relation defines the stiffness exponent y (or θ) [4,8], a fundamental quantity assessing low-temperature energy fluctuations: a positive value of y, as found in EA for $d \geq 3$, denotes the increase in the energetic cost (i.e. "stiffness") accompanying a growing number of variables perturbed from their position in the ground state. The rise in energetic penalty paid for stronger disturbances signals the presence of an ordered state. In turn, for systems with $y \leq 0$ such order is destabilized by arbitrarily small fluctuations.

But instead of determining the interface scaling on a compact lattice structure, we will focus on the interface energy $\sigma(\Delta E)$ on a bond-diluted lattice, in particular, at the percolation threshold p_c. Due to the tenuous fractal nature of the percolating cluster at p_c, no long-range order can be sustained, defects possess a vanishing interface, and [9]

$$\sigma(\Delta E)_{L,p_c} \sim L^{y_P} \quad \text{with} \quad y_P \leq 0\,. \tag{11.2}$$

Below, we will present an efficient algorithm [6] do determine y_P in (11.2) from a fit to data obtained for system sizes $L \leq 100$ in $d = 3$. The interest in this exponent stems from its relation to the "thermal-percolative cross-over exponent" ϕ defined via [9]

$$T_g(p) \sim (p - p_c)^{\phi}\,, \quad \text{with} \quad \phi = -\nu y_P\,. \tag{11.3}$$

Equation (11.3) specifies the details of the phase boundary near p_c (or p^*), which is experimentally accessible for certain materials, as shown in Fig. 11.1. A comparison between computational prediction and experiment would lend a lot of credibility to the EA and its simplifying assumptions, such as universality with respect to the details of the bond distribution $P(J)$. While we use a Gaussian bond distribution of zero mean and unit variance throughout, we need merely require that the bond distribution is continuous and finite near $P(0)$. (Discrete bonds, such as $\pm J$, with $P(J) = 0$ for an open set near $J = 0$, can be shown to yield only $\phi = y_P = 0$.)

Following the discussion in [9,11], for bond-diluted lattices at $p \to p_c$ we have to generalize the scaling relation for the defect energy $\sigma(\Delta E)$ to

$$\sigma(\Delta E)_{L,p} \sim \mathcal{Y}(p) L^y f\left(L/\xi(p)\right)\,, \tag{11.4}$$

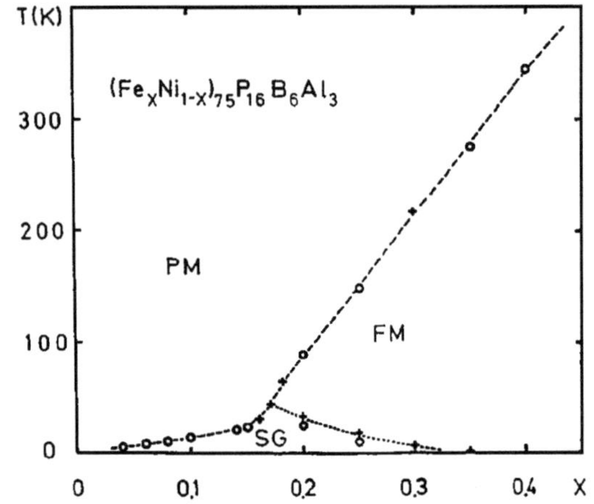

Fig. 11.1. Plot of the experimentally obtained phase diagram for $(\mathrm{Fe}_x\mathrm{Ni}_{1-x})_{75}$ $\mathrm{P}_{16}\mathrm{B}_6\mathrm{A}_3$, from [10]. Of particular interest to this study is the behavior of the phase line $T_g(x)$ near the transition $T_g(x_c) \to 0$. T_g rises near-linearly with x, according to this limited data set. Renewed experiments may significantly improve the results

where $\mathcal{Y} \sim (p - p_c)^t$ is an effective surface tension and $\xi(p) \sim (p - p_c)^{-\nu}$ is the correlation length for the cross-over into glassy behavior, which equals the corresponding percolation exponent. The scaling function $f(x)$ is defined to be constant for large argument, $L \gg \xi(p) \gg 1$.

For $\xi \gg L \gg 1$, (11.4) requires that $f(x) \sim x^\mu$ for $x \to 0$ to satisfy the vanishing of σ with L in case of Gaussian bonds. To cancel the p-dependence at $p = p_c$, (11.4) requires $y + \mu = y_P$ and $t + \mu\nu = 0$, i.e. $t = \nu y + \phi$, introducing $\phi = -\nu y_P$ in (11.3) [9].

At the cross-over $\xi \sim L$, where the range L of the excitations $\sigma(\Delta E)$ reaches the percolation length and spin glass order ensues, (11.4) yields

$$\sigma(\Delta E)_{\xi(p),p} \sim (p - p_c)^t \, \xi(p)^y f(1) \sim (p - p_c)^\phi \, . \tag{11.5}$$

Associating a temperature with this cross-over by $\beta\sigma(\Delta E)_{\xi(p),p} \sim 1$ (for temperatures above $T = 1/\beta$, thermal fluctuation destroy spin glass order), leads to the relation between p and the glass transition temperature in (11.3).

A determination of the defect energy at p_c has the advantage that the following exact reduction method [6] almost always succeeds in evaluating the ground state completely, obviating the application of any optimization and large lattice sizes can be reached.

To exploit the advantages of spin glasses on a bond-diluted lattice at p_c, we can *reduce* almost all degrees of freedom. Here, we focus exclusively on the reduction rules for the energy at $T = 0$; a subset of these also permit the exact determination of the entropy and overlap [12]. These rules apply to general Ising spin glass Hamiltonians as in (11.1) with *any* bond distribution $P(J)$, discrete or continuous, on arbitrary sparse graphs.

The reductions effect both spins and bonds, eliminating recursively at least all zero-, one-, two-, and three-connected spins. These operations eliminate and add terms to the expression for the Hamiltonian in (11.1), but leave it form-invariant. Offsets in the energy along the way are accounted for by a variable H_o, which is *exact* for a ground-state configuration.

Rule I: An isolated spin can be ignored entirely.

Rule II: A one-connected spin i can be eliminated, since its state can always be chosen in accordance with its neighboring spin j to satisfy the bond $J_{i,j}$. For its energetically most favorable state we adjust $H_o := H_o - |J_{i,j}|$ and eliminate the term $-J_{i,j}\, x_i\, x_j$ from H.

Rule III: A double bond, $J_{i,j}^{(1)}$ and $J_{i,j}^{(2)}$, between two vertices i and j can be combined to a single bond by setting $J_{i,j} = J_{i,j}^{(1)} + J_{i,j}^{(2)}$.

Rule IV: For a two-connected spin i, rewrite in (11.1)

$$\begin{aligned} x_i(J_{i,1}x_1 + J_{i,2}x_2) &\le |J_{i,1}x_1 + J_{i,2}x_2| = J_{1,2}x_1x_2 + \varDelta H\,, \qquad (11.6)\\ J_{1,2} &= \frac{1}{2}\left(|J_{i,1} + J_{i,2}| - |J_{i,1} - J_{i,2}|\right)\,,\\ \varDelta H &= \frac{1}{2}\left(|J_{i,1} + J_{i,2}| + |J_{i,1} - J_{i,2}|\right)\,, \end{aligned}$$

leaving the graph with a new bond $J_{1,2}$ between spin 1 and 2, and acquiring an offset $H_o := H_o - \varDelta H$.

Rule V: A three-connected spin i can be reduced via a "star-triangle" relation, as depicted in Fig. 11.2:

$$\begin{aligned} J_{i,1}\, x_i\, x_1 + J_{i,2}\, x_i\, x_2 + J_{i,3}\, x_i\, x_3 &\le |J_{i,1}x_1 + J_{i,2}x_2 + J_{i,3}x_3|\\ &= J_{1,2}\, x_1\, x_2 + J_{1,3}\, x_1\, x_3 + J_{2,3}\, x_2\, x_3 + \varDelta H\,, \qquad (11.7) \end{aligned}$$

where

$$\begin{aligned} J_{1,2} &= -A - B + C + D\,, \quad & J_{1,3} &= A - B + C - D\,,\\ J_{2,3} &= -A + B + C - D\,, \quad & \varDelta H &= A + B + C + D\,,\\ A &= \frac{1}{4}\left|J_{i,1} - J_{i,2} + J_{i,3}\right|\,, \quad & B &= \frac{1}{4}\left|J_{i,1} - J_{i,2} - J_{i,3}\right|\,,\\ C &= \frac{1}{4}\left|J_{i,1} + J_{i,2} + J_{i,3}\right|\,, \quad & D &= \frac{1}{4}\left|J_{i,1} + J_{i,2} - J_{i,3}\right|\,. \end{aligned}$$

The bounds in (11.6–11.7) become *exact* when the remaining graph takes on its ground state. After a recursive application of these rules, the original lattice graph is completely reduced near p_c, in which case H_o is exact.

Our studies for systems up to $L = 100$ have yielded an exponent of $y_P = -1.28(2)$. The consensus of results for the correlation exponent in $d = 3$ for percolation seems to be $\nu = 0.86(2)$ [13], which results in a thermal-to-percolative crossover exponent, see (11.3), of $\phi = 1.10(4)$. This result is quite consistent with the near-linear slope of T_g near the percolation threshold for the limited experimental data set in Fig. 11.1.

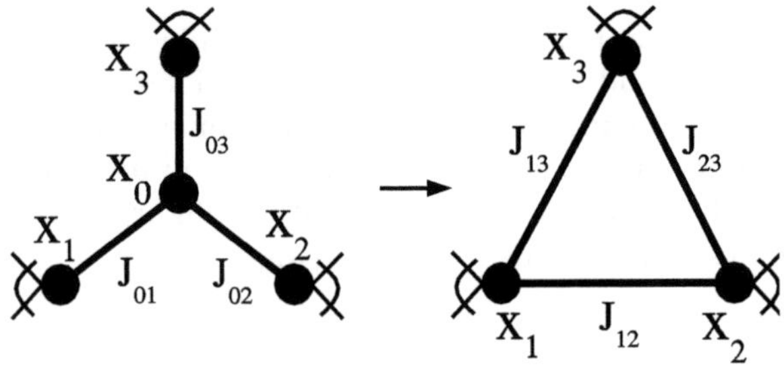

Fig. 11.2. "Star-triangle" relation to reduce a three-connected spin x_0. The new bonds on the right are obtained in (11.7)

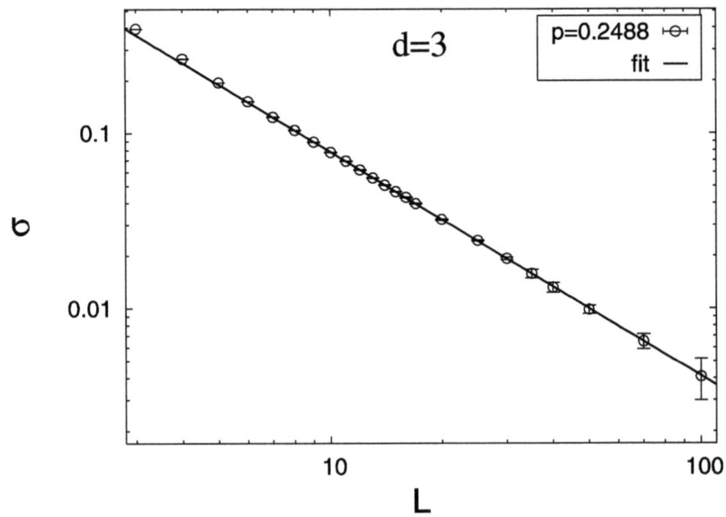

Fig. 11.3. Plot of the defect energies $\sigma(\Delta E)$ as a function of system size L in $d = 3$ at the bond-percolation threshold, according to (11.2). An asymptotic fit to the data yields $y_P = -1.28(2)$

Acknowledgement

We like to thank the organizers of the annual workshop at the Center for Simulation Physics for their kind invitation, and the NSF for financial support for this work under grant DMR-0312510.

References

1. *Spin glasses and random fields.* Ed. A.P. Young (World Scientific, Singapore 1998)
2. *New Optimization Algorithms in Physics.* Eds. A. Hartmann, H. Rieger (Springer, Berlin Heidelberg New York 2004)

3. S.F. Edwards, P.W. Anderson: J. Phys. F: Metal Phys. **5**, 965 (1975)
4. K.H. Fischer, J.A. Hertz: *Spin Glasses.* (Cambridge University Press, Cambridge 1991)
5. K. Binder, A.P. Young: Rev. Mod. Phys. **58**, 801 (1986)
6. S. Boettcher: Europhys. Lett. **67**, 453 (2004); Euro. Phys. J. B **38**, 83 (2004)
7. S. Boettcher: Phys. Rev. Lett. **95**, 197205 (2005)
8. A.J. Bray, M.A. Moore: J. Phys. C: Solid State Phys. **17**, L463 (1984)
9. J.R. Banavar, A.J. Bray, S. Feng: Phys. Rev. Lett. **58**, 1463 (1987)
10. O. Beckman, E. Figueroa, K. Gramm, L. Lundgren, K.V. Rao, H.S. Chen: Phys. Scr. **25**, 726 (1982)
11. A.J. Bray, S. Feng: Phys. Rev. B **36**, 8456 (1987)
12. S. Boettcher: Euro. Phys. J. B **33**, 439 (2003)
13. P.H.L. Martins, J.A. Plascak: Phys. Rev. E **67**, 046119 (2003)

12

Simulation Study of Gas-Liquid Interface

F. Ogushi, S. Yukawa, and N. Ito

Department of Applied Physics, School of Engineering, The University of Tokyo, 7-3-1 Hongo, Bunkyo-ku, Tokyo 113-8656, Japan

Abstract. Heat conduction of three-dimensional Lennard-Jones particle system are studied using nonequilibrium moleculer dynamics simulation. Geometry of the system is a rectanguler parallelepiped box of the size of $L_x \gg L_y \times L_z$. Two Nosé-Hoover heat bathes with different temperature $T_{H,L}$ ($T_H \geq T_L$) are attatched on the regions near both ends of x-direction. The density and $T_{H,L}$ are set to be in supercritical fluid, liquid and solid phase. In a single phase system, the heat conductivity shows the system size dependence $1/\sqrt{L_x}$ where L_x is the system size. Heat flux keeps the system in a gas-liquid coexisting state and an interface exists steadily. Its interface is thicker in gas-side than in liquid-side. Using a characteristic length $X_{g,l}$ that is desided by the constant density in each phase, we construct a minimul model of an asymmetric interface with tanh form. This model has one parameter L that is the thickness of the interface and this shows good agreement with the simulation results.

12.1 Introduction

Theoretical description of the structure of a phase interface is cruicial to understand the heat conduction of the heterogeneous system. Bubble nucleation and heat resistance are examples of them. The classical nucleation theory (CNT) [1] is a simple theory to explain the nucleation and compare with experiments. But the CNT dose not consider the finite size thickness of the interface and estimated values using the CNT do not explain experimental results. There are some approaches that aim to describe the properties of clusters on a molecular level and aim to give a diffusive interface [2–4]. For example, Oxtoby et al. modify the CNT using the density functional method [5,6]. This theory gives a diffusive interface and a nucleation rate that is more close to the experiment than the CNT.

In macroscopic scale, the heat conduction obeys the Fourier's law of the heat conduction but it dose not clear in microscopic scale. For example, some studies using $\beta-$FPU, hard-core particle system and Lennard-Jones particle system shows that the heat conductivity κ in a finite size system depends

Springer Proceedings in Physics, Volume 105
Computer Simulation Studies in Condensed-Matter Physics XVIII
Eds.: D.P. Landau, S.P. Lewis and H.-B. Schüttler

on the system size L_x [7–9]. This dependence is explained in the basis of Kubo's fomula and the long-time tail; the autocorrelation function of the heat flux $\langle j(0)j(t)\rangle$ decays with $t^{-d/2}$ where d denotes the space dimension. From Kubo's famula, κ is espressedby the time integration of $\langle j(0)j(t)\rangle$. For a finite-size system, the upper limit of the time integration is cut by a certain finite constant proportional to the system size. Therefore, finite-size correlation to κ is expected to be on the order of $L_x^{1-d/2}$.

In this article, we report the heat conduction of single phase system and the structure of gas-liquid phase interface. To give a gas-liquid phase transition, we use Lennard–Jones (12–6) potential. Some investigations on nonequilibrium phenomena using Lennard-Jones particle system, for example, the structure of the bubble [10,11] and the liquid-gas phase transition [12], have been reported. Lennard–Jones (LJ) potential between particle i and j is defined as follows:

$$\phi(r_{ij}) = 4\varepsilon \left\{ \left(\frac{\sigma}{r_{ij}}\right)^{12} - \left(\frac{\sigma}{r_{ij}}\right)^{6} \right\}, \tag{12.1}$$

where r_{ij} denotes a distance between particle i and j. The geometry of the system is a rectanguler parallelepiped box of the size of $L_x \gg L_y \times L_z$. We slice the box with fixed width $\mathrm{d}x$ in x-direction and calculate the temperature and the density in each cell. To give a temperature gradient, both end regions in x-direction are contact with Nosé-Hoover heat bathes [13,14] with different temperature T_H and T_L ($T_H \geq T_L$). The simulation details are shown in [9]. We also use next definitions of the temperature and the density. Local temperature at each local cell is calculated with $T(x) = m\langle \boldsymbol{v_i}^2\rangle_{\text{cell}}/3k_{\text{B}}$, where m, k_{B} and $\boldsymbol{v_i}$ are mass of the particle, Boltzmann constant and velocity of i-particle, respectively. Local density at each local cell is calculated with $\rho(x) = \langle n\rangle_{\text{cell}}/V$, where n and V are number of particles in the cell and the volume of each cell ($V = L_y \times L_z \times \mathrm{d}x$), respectively. $\langle \ldots \rangle_{\text{cell}}$ denotes long time average on the local cell.

12.2 Heat Conduction of A Single Phase System

The heat conduction of the homogeneous system is studied using a nonlinear lattice model and a hard-core particle system but LJ potentail has a longer range and more complex local structure than them. The size dependence may have some characteristic form and it may not be described by a simlpe $1/\sqrt{L}$. The heat conductivity $\kappa(L_x)$ is defined by

$$\kappa(L_x) = J/\frac{\mathrm{dT}}{\mathrm{d}x}, \tag{12.2}$$

where J is a heat flux and it is calculated using the work of the heat bath variables. Figure 12.1 shows the estimated values of heat conductivity in the finitesize system $\kappa(L_x)$ in supercritical fluid phase.

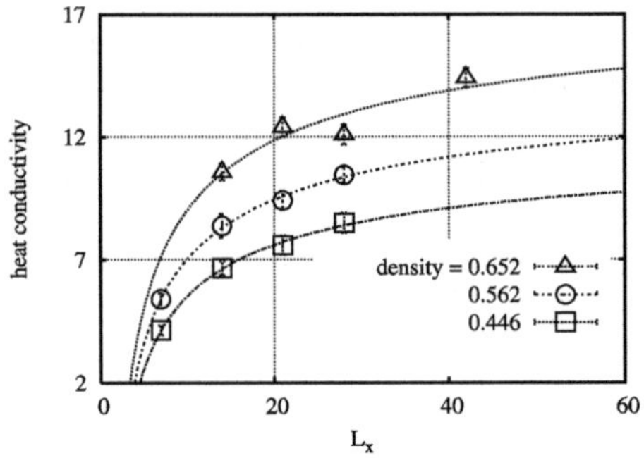

Fig. 12.1. The heat conductivity in the finite size system $\kappa(L_x)$ are plotted with the system size L_x: The system is in supercritical fluid phase. The data are fitted with $\kappa - a/\sqrt{L_x}$ ($\kappa, a \geq 0$). The system size dependence of $\kappa(L_x)$ is consistent with $1/\sqrt{L_x}$

Figure 12.1 shows that $\kappa(L_x)$ has the system size dependence $\kappa - a/\sqrt{L_x}$ where κ and a are positive. Usig this results, estimated values of the heat conductivity in macroscopic limit κ are proportional to the average density of the system. We calculate $\kappa(L_x)$ in liauid and solid phase and $\kappa(L_x)$ shows same dependence $1/\sqrt{L_x}$. The details of the size dependence of $\kappa(L_x)$ of supercritical fluid phase are shown in [9].

12.3 Gas-Liquid Coexistiong State

Using nonequiriblium molecular dynamics simulation, the multiple system is realized [8]. To realize a system with gas-liquid interface steadily, the heat bath temperature T_H, T_L and the average density ρ are set to be in a gas-liquid coexisting phase. Figure 12.2 shows the temperature-density phase diagram and the temperature changes litte even if the density changes. Thus, there is a phase interface. The parameters shown in Fig. 12.2 are different in each other.

The temperature and the density in each cell against the position decrease linearity owing to a large number of time steps applied in each phase and the values of the gradient $\mathrm{d}T/\mathrm{d}x$ and $\mathrm{d}\rho/\mathrm{d}x$ are different between each phase. The local density changes rapidly near the temperature gradient changes and there is a gas-liquid interface. The structure of the interface with a same temperature in a coexisting phase and different temperature gradient are almost same. Therefore the interface structure of our simulation is the same as one in equiriblium state.

To construct a minimul model of the asymmetric interface, we use a characteristic length $X_{g,l}$ in each phase side $\rho_{g,l} = aX_{g,l}^{-3}$, where a is a parameter and $\rho_{g,l}$ is a constant density in each phase. The density profile of the interface is described wiht $X_{g,l}$ as

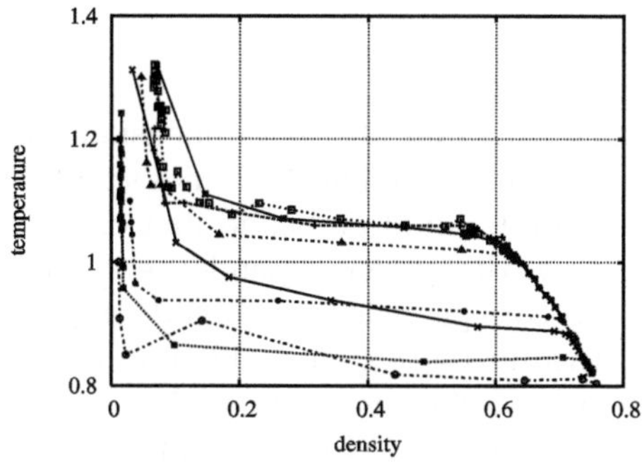

Fig. 12.2. Temperature-density phase diagram: The temperature changes little against the density in the middle area and it changes following the density in both sides. The middle area is in a gas-liquid coexisting phase and the low density side is in a gas phase and the other side is in a liquid phase

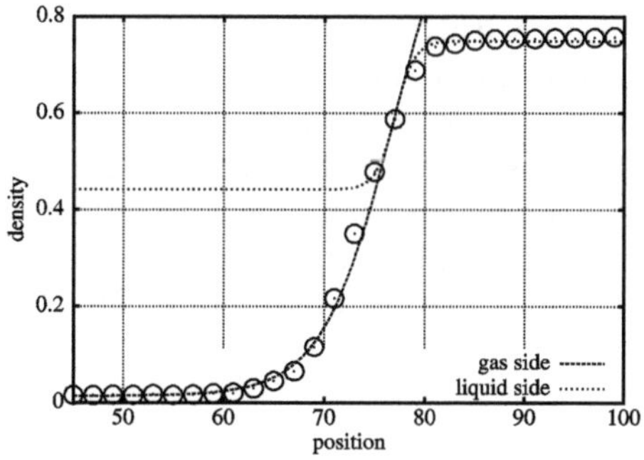

Fig. 12.3. The density profile of gas-liquid interface with $\rho_{g,l}(x)$: The model $\rho_{g,l}(x)$ agree with simulation data well. Parameters are as follows: $L_x = 130$, $L_y = L_z = 4$, $T_H = 1.32$, and $T_L = 0.8$

$$\rho_{g,l}(x) = \rho_{g,l} + \beta_{g,l}(1 - \tanh|\frac{x - x_0}{X_{g,l}}|) , \tag{12.3}$$

where $\beta_{g,l}$ is a coefficient and the x_0 is the position that devides the interface into each phase side. If we presuppose that the density changes smoothly in the interface, $\rho_{g,l}(x)$ is decided with one parameter L $(= X_g + X_g)$ that corresponds to a thickness of the interface. Figure 12.3 shows simulation data and $\rho_g(x)$ and $\rho_l(x)$. $\rho_{g,l}(x)$ shows good agreement with simulation results and this dose not depend on simulation parameters. In this model, the criticla behavior is consistent with three-dimensional Ising universality. From $\rho_{g,l} = aX_{g,l}$, the behavior of a and $L = X_g + X_l$ are $|(T - T_c)/T_c|^{-(3\nu-\beta)}$ and $|(T - T_c)/T_c|^{-\nu}$, where $\nu = 0.63$ and $\beta = 0.32$, respectively.

12.4 Summary

In summary, the heat conduction of three-dimensional LJ particle system is studied using nonequiriblium molecular dynamics simulation. In homogeneous system in supercritical fluid, liquid and solid phase, the heat conductivity in the finite size system shows the system dependence $1/\sqrt{L_x}$ and this behavior is consistent with the hard-core particle system and long-time tail. If the system has a phase interface steadily, there are two different temperature gradient and the density profile of the interface is asymmetric. To give the asymmetricity, we use a characteristic length $X_{g,l}$ that is decided with the density $\rho_{g,l}$. We presuppose that the form of the interface is tanh and the density changes smoothly in the interface. Thus, the density profile is described with one parameter L. Our model shows good agreement with simulation data and its critical behavior is consistent with the results that predicted by three-dimensional Ising model.

A free energy density with double well form gives an interface with finite size thickness. The form of the free energy with steady phase interface should have a same depth of the bottom. With a harmonic approximation, its thickness has a positive correlation with a compresibility.

From experiment, it is known that a heat resistance in a phase interface exists and this is so-called Kapitza resistance [15]. In our simulation, such temperature gap in interface was not observed in thiree-dimensional and two-dimensional system.

Acknowledgement

This work is partly supported by the Japan Society for the Promotion of Science (No. 14080204 and 14740229).

References

1. M. Volmer, H.Z. Flood: Z. Phys. Chem. A **190**, 273 (1934); R. Becker, W. Döring: Ann. Phys. **24**, 719 (1935); J. Frenkel: *Kinetic Theory of Liquids* (Theodre Steinkopff, Dresden 1939)
2. F.F. Abraham: *Homogeneous Nucleation Theory* (Academic Press, New York 1979); A. Laaksonen, V. Talanquer, D.W. Oxtoby: Annu. Rev. Phys. Chem. **46**, 489 (1995); J.M. Howe: Philos. Mag. A **74**, 761 (1996)
3. J.W. Chan, J.E. Hilliard: J. Chem. Phys. **28**, 258 (1958); J. Chem. Phys. **31**, 688 (1959)
4. F.F. Abraham: J. Chem. Phys. **60**, 246 (1974); Phys. Rep. **53**, 93 (1979)
5. D.W. Oxtoby, R. Evans: J. Chem. Phys. **89**, 7521 (1988); X.C. Zeng, D.W. Oxtoby: J. Chem. Phys. **94**, 4472 (1991); D.W. Oxtoby: J. Phys. Condensed Matter **4**, 7627 (1992)
6. K. Laasonen, S. Wonczak, R. Strey, A. Laaksonen: J. Chem. Phys. **113**, 9741 (2000); L. Gránásy, T. Pusztai, P.F. James: J. Chem. Phys. **117**, 6157 (2002)

7. T. Shimada, T. Murakami, S. Yukawa, K. Saito, N. Ito: J. Phys. Soc. Jpn. **62**, 3150 (2000)
8. T. Murakami, T. Shimada, S. Yukawa, N. Ito: J. Phys. Soc. Jpn. **72**, 1049 (2003)
9. F. Ogushi, S. Yukawa, N.Ito: J. Phys. Soc. Jpn. **74**, 827 (2005)
10. T. Kinjo, M. Matsumoto: Fluid Phase Equil. **144**, 343 (1998)
11. T. Kinjo, M. Matsumoto: Fluid Phase Equil. **14**, 138 (1999)
12. H. Okumura, D. M. Heyes: Phys. Rev. E. **70**, 061206 (2004)
13. S. Nosé: J. Chem. Phys. **81**, 511 (1984)
14. W.G. Hoover: Phys. Rev. A **31**, 1695 (1985)
15. P.L. Kapitza: J. Phys. (MOSCOW) **4**, 181 (1941)

13

Dynamic Scaling of a Cluster Growth Process far from Equilibrium

I.T. Georgiev, B. Schmittmann, and R.K.P. Zia

Department of Physics
and Center for Stochastic Processes in Science and Engineering,
Virginia Tech, Blacksburg, VA 24061-0435, USA

Abstract. The dynamics of cluster growth in a two-lane driven diffusive system is analyzed by extensive Monte Carlo simulations. We observe good dynamical scaling for small values of a particle-particle exchange probability γ on relatively small lattices. As γ increases, the (effective) scaling exponents develop a dependence on γ, suggesting that the observed coarsening does not result in a genuine ordered phase but is instead part of a complex crossover.

13.1 Introduction

The study of idealized lattice models has proved to be invaluable for exploring the vast realm of systems far from equilibrium. Although simple to define, the behavior of these models is very rich and complex [1]. Since exact solutions are only rarely available, the initial insights into their properties are almost always gleaned from direct simulations. To develop the necessary confidence in the numerics, models for which both exact solutions and simulation data can be compared have received especially close scrutiny. It is therefore particularly startling if stark discrepancies between simulation data and analytical predictions are observed. For example, the exact steady-state solution for a simple exclusion process with two species on a ring shows that a "phase transition" suggested by simulation data and mean-field theory [2] is nothing but a finite-size effect [3, 4]: Particle-particle correlations are controlled by a length scale which can become enormous (e.g., 10^{70} lattice sites) but nevertheless remains finite. Such extraordinary findings in a simple one-dimensional stochastic model reveal the unexpected, often counter-intuitive features of systems far from equilibrium.

In this work we study a closely related model, namely a system of two particle species [5, 6], driven in opposite directions on two coupled lanes [7]. When two particles encounter one another, they may exchange positions with a small exchange rate γ. Monte Carlo simulations of static and dynamic properties suggest that the system undergoes a phase transition between a homo-

Springer Proceedings in Physics, Volume 105
Computer Simulation Studies in Condensed-Matter Physics XVIII
Eds.: D.P. Landau, S.P. Lewis and H.-B. Schüttler

geneous, disordered phase and an ordered phase characterized by the presence of a single macroscopic particle cluster. Starting from an initially random configuration, this ordered phase is approached via a coarsening process in which clusters grow much faster than diffusively [8,9]. For this model, there is no exact solution which can provide rigorous insights into the nature of the ordered "phase". Yet again, it appears to be a finite-size effect, based on an analytic conjecture due to Kafri et al. [10,11] and high-precision Monte Carlo data for steady-state cluster size distributions on rather large systems [13].

In this paper we show data from extensive computer simulations focusing on the dynamic scaling properties of the system, as it approaches the steady state. For small values of γ, we find a growth exponent of 2/3 and good evidence of dynamical scaling. Yet, as γ increases, we are forced to adjust the scaling exponents in order to collapse the data on a universal curve. We believe that these findings provide further evidence for the destabilization of the ordered "phase" with increasing γ or system size.

13.2 The Model and its Properties

The model is defined as a stochastic driven diffusive lattice gas with random sequential dynamics that conserves the number of particles. Each site of a periodic $2 \times L$ lattice can be occupied by a "positive" (+) or "negative" (−) particle, or remain empty ($\oslash$). Biased in opposite directions, the two species of particles diffuse on the lattice. Their numbers are equal, i.e. the system is uncharged. The particles interact only through an excluded volume constraint. In one Monte Carlo step (MCS), the following steps are repeated $2L$ times: (i) a pair of two nearest-neighbor sites (a "bond") is chosen at random; (ii) for bonds along the L direction, a $(+\oslash)$ or a $(\oslash-)$ pair is always exchanged while a $(+-)$ pair is exchanged with probability γ; (iii) for transverse bonds, particle-hole pairs are always exchanged while particle-particle pairs are exchanged with probability γ. The chosen dynamical rules are such that the bias is from left to right (right to left) for the + (−) particles, and no backward jumps are allowed. In contrast, lane changes occur in an unbiased fashion.

Clusters of size s are defined as s particles connected by nearest-neighbor bonds regardless of their charge. To monitor the growth of such clusters quantitatively, we measure the residence distribution, $p(s,t)$, i.e., the probability that a randomly selected particle belongs to a cluster of size s at time t. It is naturally normalized, i.e. $\sum_{s=1}^{L} p(s,t) = N$, where N is the total number of particles in the system. The average cluster size is then defined by $l_c(t) = \sum_{s=1}^{L} s\, p(s,t)/N$.

In the stationary state, the residence distribution $p^*(s) = \lim_{t\to\infty} p(s,t)$ clearly distinguishes homogeneous from ordered configurations. When the system is populated only by small clusters, $p^*(s)$ is monotonically decreasing (unimodal) in s. A bi-modal $p^*(s)$ signals the presence of a macroscopic cluster,

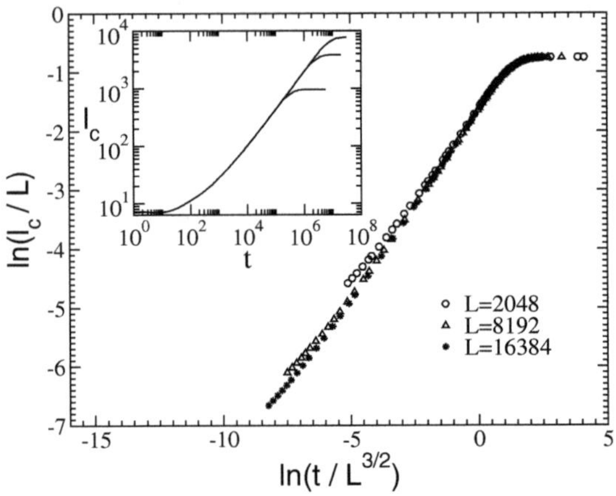

Fig. 13.1. Dynamic scaling for l_c/L^α vs. t/L^z for $\gamma = 0.1$ and $\rho = 0.5$. The inset shows the original data l_c vs. t

coexisting with a low-density ("traveller") region in the remainder of the system [13]. Our steady-state data show clearly that bi-modal distributions cross over to uni-modal ones, if either γ or L are increased [13]. In contrast, one-lane systems exhibit exponentially decaying (uni-modal) distributions over the whole parameter space, in agreement with an exact solution [15].

13.3 Monte Carlo Simulations

Faced with the computational challenge to simulate a slow diffusion process on large lattices, we apply the "brute force" method. To improve the statistics of the simulations, we use a fast multi-spin coding algorithm for conserved dynamics (see, e.g., [12]). Our model is a three-state model so that we need at least two bits to store the state of a particular site on the lattice. Instead of encoding such a state by two consecutive bits in one machine word, we prefer to use one bit from two different words. In this way the single-bitwise algorithm for one Monte Carlo step becomes a little bit more efficient, but we suffer a (mild) speed penalty from working with two different words for the encoding. The multi-spin code that we use for a single bond exchange is more complicated than standard codes but, using hardware with 64-bit architecture, we benefit much more from simulating 64 lattices at the same time. The overall efficiency increases by about a factor of 35.

The inset of Fig. 13.1 shows l_c vs. t obtained from 4096 independent runs. Data are collected for lattices with $L = 2\text{K}$, 8K and 16K at particle density $\rho = 0.5$ and $\gamma = 0.1$. At $t = 0$ the initial configuration on the lattice is random. We test whether the data obey dynamic scaling of the form

$$l_c(t) = L^\alpha\, F(t/L^z)\,. \tag{13.1}$$

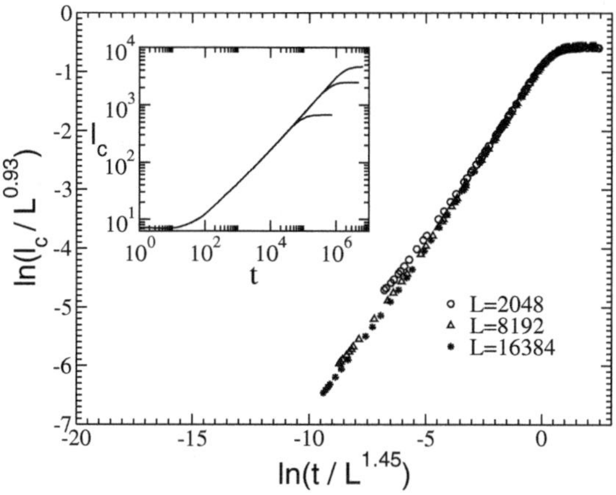

Fig. 13.2. Dynamic scaling for l_c/L^α vs. t/L^z for $\gamma = 0.3$ and $\rho = 0.5$. The inset shows the original data l_c vs. t

We obtain very good data collapse for large t ($t > 1000$ MCS), using the values $\alpha = 1.0$ and $z = 3/2$ for the dynamic exponent (see Fig. 13.1). This results in a cluster growth exponent $\beta = \alpha/z = 2/3$ with error bars much smaller than the symbols. For this small value of γ and for the L's that we could simulate, the growth exponent β appears to coincide with the prediction from the Burgers equation [14]. Since the early-time behavior of $l_c(t)$ is essentially independent of L, as illustrated by the inset of Fig. 13.1, the scaling form of (13.1) cannot describe the full range of the data. The plateau in $l_c(t)$ at early times can be understood easily: starting from a random configuration, small clusters form almost immediately everywhere on the lattice, as oppositely charged particles block each other. Until particles make their way through these clusters and begin to escape from the opposite side, $l_c(t)$ cannot change significantly.

For this small value of $\gamma = 0.1$, the dynamic scaling exponents are consistent with the formation of a single macroscopic cluster whose size, in steady state, scales with the system size. However, a different picture emerges for larger values of γ. The data for $\gamma = 0.3$ are shown in Fig. 13.2. Here, a least squares minimization gives the best scaling collapse if $\alpha \approx 0.93(1)$ and $z \approx 1.45(1)$, whence $\beta \approx 0.64(1)$. It is tempting to conjecture that the apparent γ-dependence of the scaling exponents is evidence for the finite-size nature of the ordered "phase". By increasing γ, we effectively destabilize the macroscopic cluster, which in turn produces the shift in the cluster growth exponents.

13.4 Conclusions

In this paper, we have discussed the dynamic scaling properties of coarsening particle clusters in a two-species driven diffusive system on two coupled lanes. We find excellent data collapse with Burgers scaling exponents at small values of γ (0.1); in contrast, at larger γ (0.3), good data collapse can only be achieved with a different set of (effective) exponents. We believe that this is a *dynamic precursor* of dramatic differences in the steady states at these parameter values. While systems at small γ (for the L's investigated) are characterized by a single macroscopic cluster in the long-time limit, systems at larger γ can already sustain several large clusters which, however, do no longer scale well with the system size. These findings serve as another indicator that the observed ordered states do not constitute a genuine thermodynamic phase.

Acknowledgments

We thank J.T. Mettetal, G. Prüssner and M. Mobilia for stimulating discussions. This research is supported in part by the US National Science Foundation through grant DMR-0414122.

References

1. B. Schmittmann, R.K.P. Zia: Statistical mechanics of driven diffusive systems. In: *Phase Transitions and Critical Phenomena*, Vol. 17, ed. by C. Domb and J.L. Lebowitz (Academic Press, New York 1995)
2. P.F. Arndt, T. Heinzel, V. Rittenberg: J. Phys. A: Math. Gen. **31**, L45 (1998)
3. N. Rajewsky, T. Sasamoto, E.R. Speer: Physica A **279**, 123 (2000)
4. T. Sasamoto, D. Zagier: J. Phys. A: Math. Gen. **34**, 5033 (2001)
5. B. Schmittmann, K. Hwang, R.K.P. Zia: Europhysics Letters **19**, 19 (1992)
6. G. Korniss, B. Schmittmann, R.K.P. Zia: Europhysics Letters **32**, 49 (1995)
7. G. Korniss, B. Schmittmann, R.K.P. Zia: Europhys. Lett. **45**, 431 (1999)
8. J.T. Mettetal, B. Schmittmann, R.K.P. Zia: Europhys. Lett. **58**, 653 (2002)
9. B. Schmittmann, J.T. Mettetal, R.K.P. Zia, in *Computer Simulation Studies in Condensed Matter Physics XVI.* Eds. D.P. Landau, S.P. Lewis, H.B. Schüttler (Springer, Berlin Heidelberg New York 2004)
10. Y. Kafri, E. Levine, D. Mukamel, G.M. Schütz, J. Török: Phys. Rev. Lett. **89**, 035702 (2002)
11. Y. Kafri, E. Levine, D. Mukamel, J. Török: J. Phys. A: Math. Gen. **35**, L459 (2002)
12. M.E.J. Newman, G.T. Barkema: *Monte Carlo Methods in Statistical Physics.* (Oxford University Press 1999)
13. I.T. Georgiev, B. Schmittmann, R.K.P. Zia: Phys. Rev. Lett. (in press, cond-mat/0502209)
14. J.M. Burgers: *The nonlinear diffusion equation.* (D. Reidel, Dordrecht 1974)
15. S. Sandow, C. Godrèche, unpublished (1998), see also [10, 11]

14

Analysis of the Blume–Capel Model with the Wang–Landau Algorithm

D. Hurt[1], M. Eitzel[2], R. Scalettar[1], and G. Batrouni[3]

[1] Department of Physics, University of California, Davis CA 95616, USA
[2] Department of Geology, University of California, Santa Barbara CA 93106, USA
[3] Institut Non-Linéaire de Nice Université de Nice–Sophia Antipolis, 1361 route des Lucioles, 06560 Valbonne, France

Abstract. We use the Wang–Landau algorithm to investigate the thermodynamic properties of the two-dimensional ferromagnetic Blume–Capel (BC) Model on a square lattice near the tricritical point. In this region, the energy levels of the BC Hamiltonian have a much greater density than for the Ising model. Together with the necessity of distinguishing first and second order transitions, the BC Hamiltonian thus poses a challenging test of the effectiveness of the Wang–Landau method.

14.1 Introduction

The Blume–Capel Model [1,2] (BC) and its generalization, the Blume–Emery–Griffiths model [3] (BEG) have been proposed to describe phase separation and superfluidity of ^{3}He-^{4}He mixtures. They are also of intrinsic interest in the study of critical phenomena, since their phase diagram has both first-order and second-order transition lines and a tricritical point [4]. The BEG Hamiltonian is

$$\mathcal{H} = -J \sum_{\langle i,j \rangle} \sigma_i \sigma_j - K \sum_{\langle i,j \rangle} \sigma_i^2 \sigma_j^2 + \Delta \sum_i \sigma_i^2 .$$

Here J is a ferromagnetic coupling between near neighbor spins, σ_i, which can take on three possible values, $\sigma_i = \pm 1, 0$. Δ is the crystal-field coupling [5] which controls the density of vacancies, ($\sigma_i = 0$). We will focus on the more simple BC case, in which $K = 0$. For $\Delta \to -\infty$, vacancies are suppressed, and the BC model maps onto the Ising model. The ferromagnetic transition is thus expected to be second order. On the other hand, at $T = 0$, there is a first order transition at $\Delta = 2J$ between a fully spin polarized and a completely vacant lattice. A mean field treatment shows that, for a square lattice, these limits are separated by a tricritical point at $T_t = 4J/3$ and $\Delta_t = 8 \ln 2 \, J/3 = 1.848J$.

There have been a number of previous studies of this model in $d = 2$. Burkhardt used a position-space renormalization-group approach and found

Springer Proceedings in Physics, Volume 105
Computer Simulation Studies in Condensed-Matter Physics XVIII
Eds.: D.P. Landau, S.P. Lewis and H.-B. Schüttler

the tricritical temperature, $T_t/J = 0.580$ and Delta, $\Delta_t/J = 1.972$ [7]. Landau and Swendsen, using the Monte Carlo renormalization group calculated $T_t/J = 0.67$ and $\Delta_t/J = 1.94$ [8], and later a more precise value $T_t/J = 0.6091 \pm 0.0030$ and $\Delta_t/J = 1.9655 \pm 0.0151$ [9]. Selke and Yeomans [10] examined the interface properties of the model using Monte Carlo (MC) techniques to obtain $\Delta_t/J = 1.96$–1.98. Beale, using phenomenological finite-size scaling calculated the tricritical point at $T_t/J = 0.610 \pm 0.005$ and $\Delta_t/J = 1.9655 \pm 0.001$ [5]. All these critical temperatures are, as expected, considerably reduced from the mean-field value.

The goal of this paper is to examine the thermodynamic proprieties of the BC model and look at the effectiveness of the Wang–Landau algorithm in locating the tricritical point.

14.2 Computational Approach

Recently, Wang and Landau (WL) [6] introduced a MC algorithm to achieve broad phase space sampling and evaluate the density of states $g(E)$. The calculation of $g(E)$ proceeds in stages. One begins with some initial $g(E)$, for example a constant, and suggests single spin-flip moves which are accepted with probability:

$$p(E_1 \to E_2) = min \left[\frac{g(E_1)}{g(E_2)}, 1 \right] .$$

Here E_1 and E_2 are the energies of the system before and after the move, respectively. Every time an energy state E is visited, $g(E)$ is updated by a factor f_i. Stage i of the simulation is complete once the energy histogram reaches a sufficient flatness. At that point, the visitation distribution is reset to zero, and the modification factor f_i is updated according to $f_{i+1} = \sqrt{f_i}$. The entire simulation finishes when f_i reaches a value close enough to 1.

The Wang-Landau algorithm has seen many successful applications recently. As for all new methods, it is useful to explore performance in more and more challenging situations [11]. The BC model constitutes one such difficult test case. One reason is the tricritical point, since distinguishing first and second order transitions is often difficult. In addition, whereas for the Ising model the energy levels are integers, in the BC model at general (noninteger) values of Δ, the energy levels are much more dense.

We study lattice sizes up to $N = 32 \times 32$, with a flatness requirement of 0.95 for the termination of a stage, and a limiting modification factor $\log(f_i) = 10^{-9}$. The simulations were checked for flatness every 10^6 MC sweeps, and finished in 10^9 to 10^{10} sweeps. This is significantly longer than was necessary for the models that were originally analyzed [6], and presumably reflects the more complex character of the BC model energy landscape. We focus our attention on values of Δ near $2J$ where the tricritical point is located.

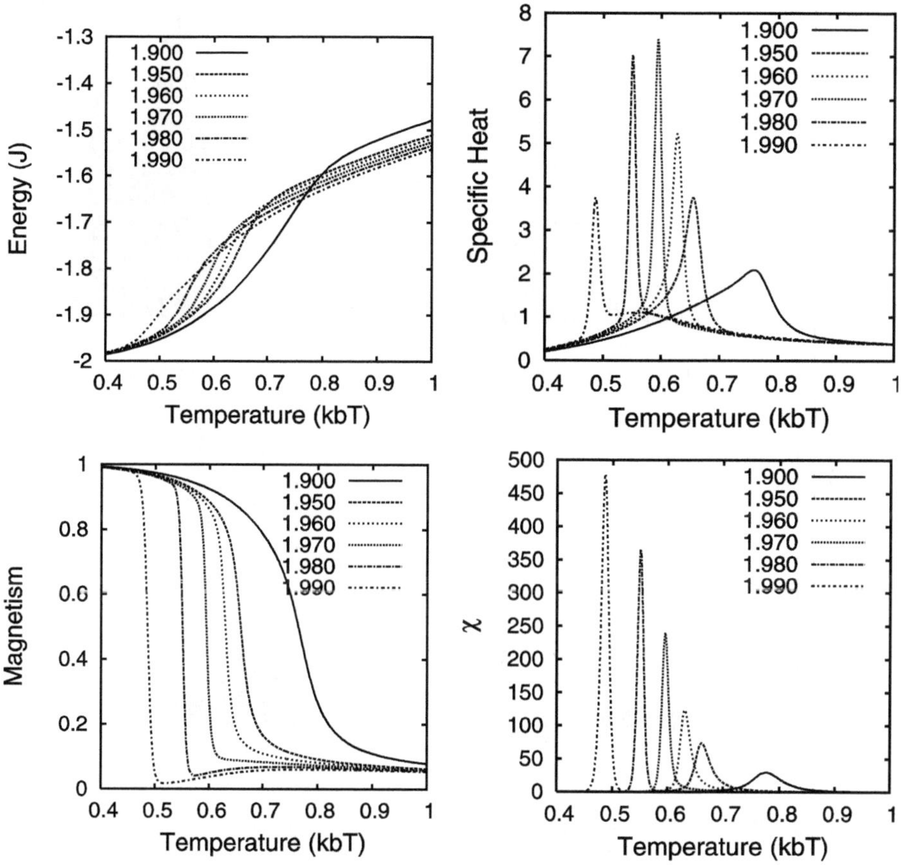

Fig. 14.1. Energy (*top left*), Specific Heat (*top right*), Magnetization (*bottom left*) and Magnetic Susceptibility (*bottom right*) vs. Temperature for $N = 32 \times 32$ and different Δ

14.3 Results

In Fig. 14.1 we show the energy and specific heat versus temperature for different values of Δ on 32×32 lattices. The peaks give clear indications of the location of the magnetic phase transition. The value of T_c at $\Delta = 1.97J$ is $T_c \approx 0.56J$.

In principle, one could examine the scaling of the specific heat maximum as described by WL [6] and others [12, 13] in order to show the transition is first order. However, this method becomes difficult for weakly first order transitions. In a study of the $q = 5$ state Potts model, Landau shows a lattice size of is required for the first order nature of the scaling of the specific heat to become apparent. We were not able to simulate large enough lattices to apply this approach to the BC model.

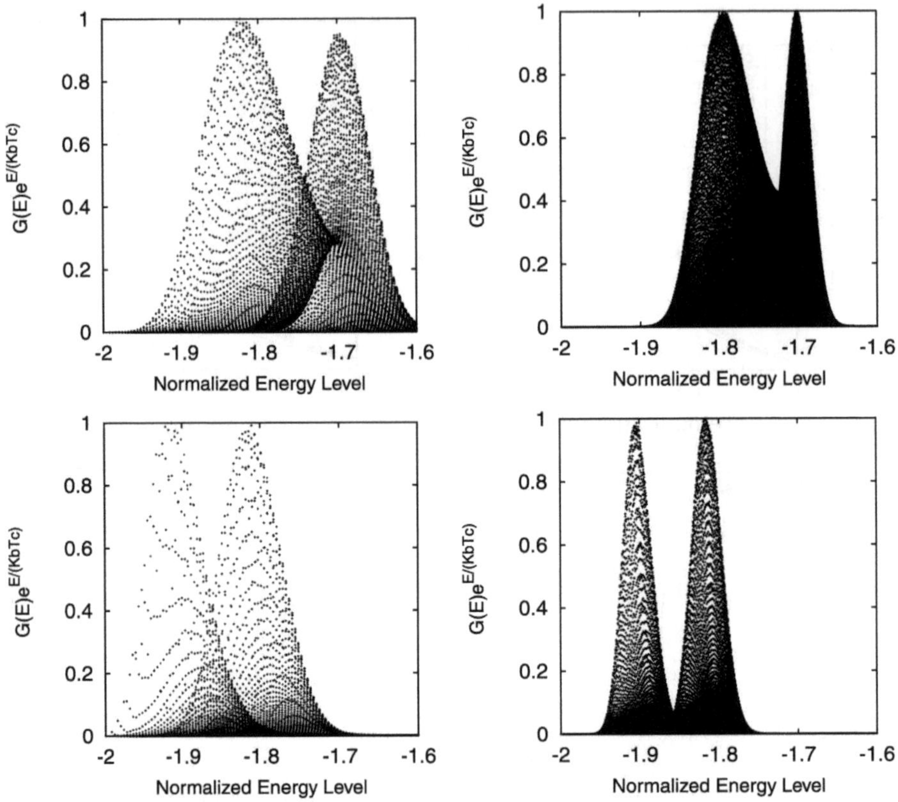

Fig. 14.2. Plots of the canonical distribution for $\Delta = 1.960J$ on top and $\Delta = 1.980J$ on the bottom. *Left*: $L = 16$ and $T_c = 0.5670J$. *Right*: $L = 32$ and $T_c = 0.5527J$

In Fig. 14.1 we examine the magnetization and magnetic susceptibility $\chi = \beta[\langle M^2\rangle - \langle M\rangle^2]$ as a function of temperature. It is clear that the relatively continuous behavior at $\Delta = 1.90J$ is replaced by a more abrupt evolution as Δ increases. This suggests that a change in the nature of the transition may be occurring.

To examine the order of the transition more rigorously, we compute the canonical distribution of the energy at $T = T_c$. At a first order transition, this should exhibit a double peak structure, where the depth of the minimum between the peaks increases with increasing lattice size. Figure 14.2 on the top shows the canonical distribution at $\Delta = 1.960J$ on 16×16 and 32×32 lattices. The depth of the trough between the double peak structure does not increase noticeably with system size, and hence this point is in the second order region. In Fig. 14.2 on the bottom we show results at $\Delta = 1.980J$. Now, with increasing system size, the trough between the double peak structure grows markedly. The transition is first order.

Figure 14.2 suggests that the tricritical point of the BC model lies between $\Delta = 1.96J$ and $\Delta = 1.98J$. Analysis of a series of such figures leads us to the conclusion that $\Delta_t/J = 1.97 \pm 0.01$. Returning now to Fig. 14.1, we see that the maximum value of the specific heat maxima occurs at $\Delta = 1.97J$.

14.4 Conclusions

In this paper we have examined the effectiveness of the Wang–Landau algorithm in determining the tricritical point of the Blume–Capel model. We have found $\Delta_t/J = 1.97 \pm 0.01$.

We did, however, encounter a number of challenges in applying the WL algorithm to this model, which we believe are linked to the high density of allowed enegy levels and the presence of the tricritical point itself. Wang and Landau [6] mention that it takes on the order of 10^7 visits for the algorithm to converge. So when the number of levels becomes large, convergence is more difficult to obtain. An obvious solution to this problem is to attempt to bin multiple energy levels together. However, we found the algorithm had difficulty converging to a flat visitation histogram. We believe the reason is that the energy levels that are close together, and form a single energy bin, may not be from lattice configurations that are close together, in the sense of the energy levels which are traversed by single spin flip moves between them. So to get some of the energy levels, the algorithm has to visit lattice configurations that are not very probable, but are lumped into a energy bin with configurations that are much more probable. Thus the simulation cannot access all the energy levels very efficiently.

To reduce the number of energy levels that the simulation had to visit, we also examined the effectiveness of dividing the energy range into windows that could be examined independently. Evenly dividing the region did not reduce the simulation time noticeably and led to fluctuations in the thermodynamic curves, presumably from where the energy windows were joined. This problem was noted with the algorithm explored by Schulz, Binder and Müller [14].

Acknowledgements

This work was supported by NSF–ITR–0313390.

References

1. M. Blume: Phys. Rev. **141**, 517 (1966)
2. H.W. Capel: Physica **32**, 966 (1966)
3. M. Blume, V.J. Emery, R.B. Griffiths: Phys. Rev. A **4**, 1070 (1971)
4. F.C. Alcaraz, J.R. Drugowich de Felício, R. Köberle, J.F. Stilck: Phys. Rev. B **32**, 7469 (1985)

5. P.D. Beale: Phys. Rev. B **33**, 1717 (1986)
6. F. Wang, D.P. Landau: Phys. Rev. Lett. **86**, 2050 (2001); Phys. Rev. E **64**, 056101 (2001)
7. T.W. Burkhardt: Phys. Rev. B **14**, 1196, (1976)
8. D.P. Landau, R.H. Swendsen: Phys. Rev. Lett. **46**, 1437 (1981)
9. D.P. Landau, R.H. Swendsen: Phys. Rev. B **33**, 7700 (1986)
10. W. Selke, J. Yeomans: J. Phys. A **16**, 2789 (1983)
11. Q. Yan, J. de Pablo: Phys. Rev. Lett. **90**, 035701 (2003)
12. K. Binder, in *Finite–Size Scaling and Numerical Simulation of Statistical Systems*, ed. by V. Privman (World Scientific, Singapore 1990), p. 173
13. D.P. Landau, in *Finite-Size Scaling and Numerical Simulation of Statistical Systems*, ed. by V. Privman (World Scientific, Singapore 1990); p. 223.
14. B.J. Schulz, K. Binder, M. Müller: Int. J. Mod. Phys. C **13**, 477, (2002)

15

Aging at Criticality in Models with Absorbing States

J.J. Ramasco[1], M. Henkel[2], M.A. Santos[3], and C.A. da Silva Santos[3]

[1] Physics Department, Emory University, Atlanta GA 30322, USA
[2] Laboratoire de Physique des Matériaux, U. Henri Poincaré Nancy I and CNRS, B.P. 239, F – 54506 Vandœuvre-lès-Nancy Cedex, France
[3] Departamento de Física and Centro de Física do Porto, Faculdade de Ciências, Universidade do Porto, Rua do Campo Alegre 687, P–4169-007 Porto, Portugal

Abstract. We study the behavior of the two time correlation functions and of the response function to an external field of the Contact Process, a paradigm model in the Directed Percolation (DP) universality class. In particular, we find that the autocorrelation and autoresponse exponents λ_Γ and λ_R are equal but, in contrast to systems relaxing to equilibrium, the ageing exponents a and b are distinct. A recent proposal to define a non-equilibrium temperature through the short-time limit of the fluctuation-dissipation ratio is therefore not applicable.

15.1 Introduction

The dynamics of glassy systems shows a set of fascinating peculiarities that have motivated a big deal of attention by part of the statistical physics community during the last decade. Apart from the very slow dynamics characteristic of glasses, these systems may also present aging, memory and rejuvenation effects (see [1–3] for recent reviews). The droplet model provides a specially simple theoretical framework to understand all these phenomena. The main idea behind this model is the existence of correlated domains of average size $\ell(t)$ in the system. To visualize these domains, one can think of the spin clusters in a critical Ising model. The size of the correlated domains grows extremely slowly in time and its growth rate crucially depends on some external parameters, as for example, the temperature.

This theoretical scheme may be further simplified if one focuses on critical systems rather than on the more complex full glasses. Critical systems also exhibit slow dynamics; in fact when a model approaches a critical point suffers the so called critical slowing down effect. In addition, the droplet model is exact at the critical points: the size of the correlation length grows then as a power law with time, $\ell(t) \sim t^{-1/z}$, and together with the system size is the only characteristic scale. This situation leads to a scaling form for most of the

Springer Proceedings in Physics, Volume 105
Computer Simulation Studies in Condensed-Matter Physics XVIII
Eds.: D.P. Landau, S.P. Lewis and H.-B. Schüttler

magnitudes characterizing critical system including those accounting for the dynamics. Two of these quantities are the two time correlation function and the response function. The two time correlation function is defined as

$$\Gamma(t,s) = \overline{\langle (\phi(t,\boldsymbol{r}) - \overline{\phi}(t))\,(\phi(s,\boldsymbol{r}) - \overline{\phi}(s)) \rangle}\,, \tag{15.1}$$

where $\phi(t, \boldsymbol{r})$ stands for the main microscopic variable of the model, e.g. the local value of the spin in an Ising model, s $(< t)$ will receive from now on the name of waiting time, $\overline{a}$ represents the average over the space of the variable a and $\langle a \rangle$ the average over disorder realization. In parallel, the response function to an external perturbation h introduced at the time s in the site $\boldsymbol{r}$ is expressed as

$$R(t,s) = \overline{\left\langle \frac{\delta\phi(t,\boldsymbol{r})}{\delta h(s,\boldsymbol{r})} \right\rangle}\,. \tag{15.2}$$

At criticality, these magnitudes follow the scaling ansatz

$$\Gamma(t,s) \sim s^{-b} f_\Gamma(t/s)\,, \quad R(t,s) \sim s^{-1-a} f_R(t/s)\,, \tag{15.3}$$

where the functions $f_{\Gamma,R}(y)$ have the following asymptotic behavior for $y \to \infty$

$$f_\Gamma(y) \sim y^{-\lambda_\Gamma/z}\,, \quad f_R(y) \sim y^{-\lambda_R/z}\,. \tag{15.4}$$

Here λ_Γ and λ_R are called the autocorrelation and autoresponse exponents, respectively.

The importance of the two time correlation and the response function comes from the fact that they are tied by the Fluctuation Dissipation relation (FDR):

$$R(t,s) = \frac{X}{T} \frac{\partial \Gamma(t,s)}{\partial s}\,. \tag{15.5}$$

This expression is a generalization for dynamical systems of the well known Fluctuation Dissipation theorem of equilibrium. The FDR also supplies a way to measure a temperature (if $X = 1$) or an effective temperature ($T_{\text{eff}} = T/X$) otherwise.

Recently, Dornic et al. [4] have shown that a temperature may be defined using the relation (15.5) by a set of critical systems including the voter model. That systems like the Ising model has a temperature, which coincides with the external parameter T by self-consistency, is not surprising. However, the same cannot be expected for the voter model that is a generic non equilibrium model (it does not fulfilled the detailedbalance condition). This is the reason why in this work we address the question of whether the same scheme is valid or not for the non equilibrium systems in general starting by the Contact Process; a paradigmatic model in the Directed Percolation universality class, the most important class of phase transitions with absorbing states.

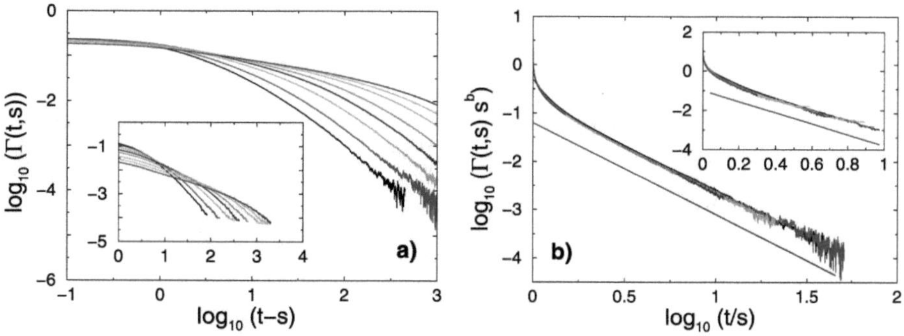

Fig. 15.1. (**a**) Evolution of the connected autocorrelation function at criticality. The curves in the inset are for two dimensions and the main plots for $1D$. The waiting times are in both cases $s = [10, 20, 50, 100, 200, 500, 1000, 2000]$ from left to right. In (**b**) the collapse of the previous curves using (15.3) and (15.4) is shown. The straight lines yield exponent values of $\lambda_\Gamma/z = 1.9$ in $1D$ and of $\lambda_\Gamma/z = 2.8$ in $2D$

15.2 The Model

The contact process (CP) was originally conceived as a simple model to describe epidemic disease propagation. In the contact process, the states of the system are described by a discrete variable, $n_i(t)$, defined on the sites i of a hypercubic lattice. The possible values of n_i are 1 or 0 depending on whether the site i is occupied or empty. The dynamics is defined as follows: for each time step, a site i of the lattice is randomly selected. If i is occupied, that particle vanishes with probability p. Otherwise, with probability $1-p$, a new particle is created on one of the nearest neighbors of i chosen at random (if that new site was still empty). When the control parameter p is varied, the model goes through a continuous phase transition from an active phase, where the mean density $\langle n(t) \rangle$ tends to a constant value $\overline{n}$ in the stationary state, to an absorbing phase with zero final density. Separating these two phases, there is a critical point located at $p_c = 0.2326746(5)$ in $1D$ (the number in brackets gives the uncertainty in the last digit), and at $p_c = 0.37753(1)$ in $2D$. The introduction of an external field in the model will be necessary to estimate a response function. For this propose, the external field h corresponds to an additional probability of creation of particles.

15.3 Results

15.3.1 Two Time Correlation Function

Similarly to what happens in the magnetic systems at the critical point, the two time correlations show aging for the CP. As may be seen in Fig. 15.1,

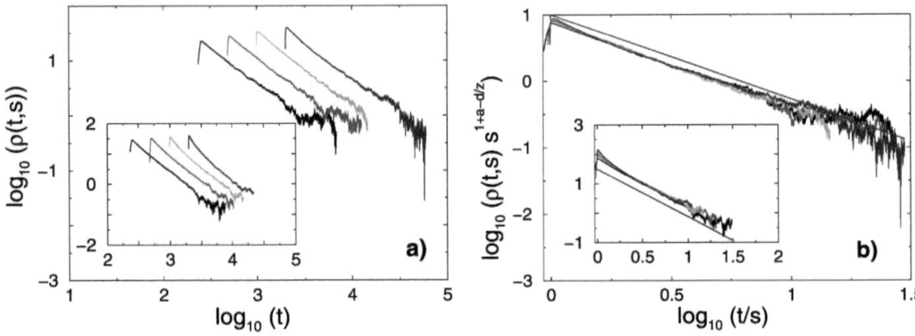

Fig. 15.2. (**a**) Temporal evolution of the 'thermoremanent' density $\rho(t,s)$ (see (15.6)) for several waiting times, from bottom to top $s = [250, 500, 1000, 2000]$. The main plots are for one dimensional systems and the insets for $2D$. In (**b**), we collapse the curves of (**a**) according to dynamical scaling (15.6). The slopes of the straight lines are -1.27 in $1D$ and -1.62 in $2D$

the curves for $\Gamma(t,s)$ with different waiting times do not overlap. However, they can all be collapsed following the scaling ansatz of (15.3) and (15.4). The value of the exponents b and λ_Γ/z turn out to be

$$b = 2\,\delta\,, \quad \lambda_\Gamma/z(1d) = 1.9 \pm 0.1 \quad \text{and} \quad \lambda_\Gamma/z(2d) = 2.8 \pm 0.3\,,$$

where δ is the exponent for the decay of the density of particles with time at criticality, $\langle \overline{n} \rangle \sim t^{-\delta}$.

15.3.2 Response Function

Due to numerical difficulties, instead of measuring the response function directly we estimate the *thermo-remanent density* $\rho(t,s)$. Both quantities are related by the expression

$$\rho(t,s) = \int_{s-\tau}^{s} dy\, R(t,y)\,,$$

and the scaling of ρ given that $R(t,s)$ follows (15.3) and (15.4), that τ is small enough compared with s and in the limit of big values of t and s is

$$\rho(t,s) = s^{-1-a+d/z} \left(\frac{t}{s}\right)^{(d-\lambda_R)/z} . \tag{15.6}$$

In Fig. 15.2, the asymptotic behavior of $\rho(t,s)$ is displayed for several waiting times. From the collapse of these curves, we get the values of the exponents

$$a = 2\,\delta - 1\,, \quad \lambda_R/z(1d) = 1.9 \pm 0.1 \quad \text{and} \quad \lambda_R/z(2d) = 2.75 \pm 0.1\,.$$

15.4 Conclusions

The fact that $a \neq b$ implies that the fluctuation-dissipation relation does not hod for the models in DP universality class. The way in which the relation is broken is a very particular one since implies that $R(t, s)$ and Γ follow the same scaling and may be thus considered as proportional. To further detail on this result see [5,6].

References

1. L.F. Cugliandolo, in *Slow Relaxation and non equilibrium dynamics in condensed matter*, Ed. J.-L. Barrat, J. Dalibard, J. Kurchan, M.V. Feigelman. (Springer, Berlin Heidelberg New York 2003). See also cond-mat/0210312 for a full review on the field.
2. C. Godrèche, J.-M. Luck: J. Phys. Cond. Matt. **14**, 1589 (2002)
3. M. Henkel: Adv. Solid State Phys. **44**, (2004) in press, cond-mat/0404016
4. I. Dornic, H. Chaté, J. Chave, H. Hinrichsen: Phys. Rev. Lett. **87**, 045701 (2001)
5. J.J. Ramasco, M. Henkel, M.A. Santos, C.A. da Silva Santos: J. Phys. A **37**, 10497 (2004)
6. T. Enss, M. Henkel, A. Picone, U. Schollwöck: J. Phys. A **37**, 10479 (2004)

16

Size Dispersity Effects on the Two-Dimensional Melting

H. Watanabe[1], S. Yukawa[2], and N. Ito[2]

[1] Department of Complex Systems Science, Graduate School of Information Science, Nagoya University, Furouchou, Chikusaku, Nagoya 464-8601, Japan
[2] Department of Applied Physics, School of Engineering, The University of Tokyo, Hongo, Bunkyo-ku, Tokyo 113-8656, Japan

16.1 Introduction

Recently, there has been considerable interest in the melting behavior of systems with size dispersity, i.e., these systems in which the radii of constituent particles are not all identical [1, 2]. The effect of the size dispersity was first treated by Dickinson and Parker [3]. They investigated the phase diagram of a model colloidal dispersion. Vermölen and Ito studied the pressure-density diagram of elastic-disk systems and found that the intermediate phase between the solid and the fluid phase vanishes above a critical dispersity [4]. Sadr-Lahijany et al. studied the density-dispersity phase diagram of Lennard-Jones particle systems [5] and found the multi-critical point at which the melting transition becomes the first order from the Kosterlitz–Thouless type. Recently, the melting behavior of the hard-disk system was studied by the Non-equilibrium relaxation method [6–9]. The critical point ρ_i between the hexatic and isotropic phases and the critical exponent were determined by observing the fluctuation of the order parameter [10, 11]. This scaling technique for determining the melting point can be applied directly to the system with dispersity.

16.2 The Closest Packing Density

In this work, hard-disk systems with equimolar bi-dispersity (two kinds of particles of the same number) are treated. The density of the system is defined to be $\rho = 4\sum_i^N r_i^2/A$, with the number of particles N, a radius r_i and the area of the system A. The dispersity σ is defined by the ratio of the standard deviation to the average radius. For the equimolar bi-disperse system, $\sigma = (r_a - r_b)/(r_a + r_b)$ with two radii r_a and r_b ($r_a > r_b$).

Springer Proceedings in Physics, Volume 105
Computer Simulation Studies in Condensed-Matter Physics XVIII
Eds.: D.P. Landau, S.P. Lewis and H.-B. Schüttler

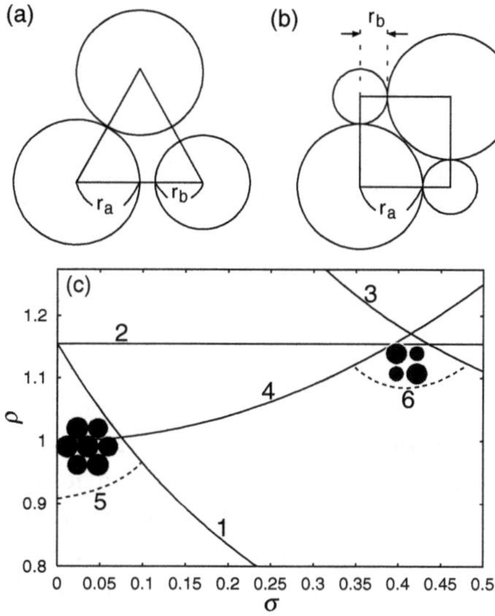

Fig. 16.1. Structures of the bi-disperse system of (**a**) Hexagonal structure and (**b**) Quadratic structure. (**c**) Predicted phase boundaries of the equimolar bi-disperse systems. Units of all plots in this paper are dimensionless. *Solid lines*: *1.* Upper density limit of the hexagonal structure $\rho_\mathrm{h}(\sigma)$ in (16.2). *2.* The highest density of the hexagonal packing $\rho = 2/\sqrt{3}$. *3.* A condition for highest density of the quadratic packing in (16.3). *4.* Another condition for the quadratic packing in (16.4). *Dashed lines* are expected lower boundaries of the hexatic *5* and of the *quadratic solid 6.* Other possible structures are not shown here

16.2.1 Hexagonal Packing

The closest hexagonal packing with the dispersity is achieved when the larger particles are just in contact (see Fig. 16.1a). The closest density of the hexagonal packing ρ_h is expected to be

$$\rho_\mathrm{h} = \frac{r_a^2 + r_b^2}{\sqrt{3} r_a^2} \,. \tag{16.1}$$

Equation (16.1) is also expressed to be

$$\rho_\mathrm{h}(\sigma) = \frac{2}{\sqrt{3}} \frac{\sigma^2 + 1}{(\sigma + 1)^2} \,. \tag{16.2}$$

While (16.2) is a simply decreasing function to dispersity σ, the melting point will shift to a higher density with a larger dispersity since it becomes more difficult to maintain a regular structure with dispersity. Therefore, the melting point will disappear in the dispersity-density parameter space when the melting density becomes higher than the highest density of the hexagonal structure.

16.2.2 Quadratic Packing

For the case of the bi-disperse system, the regular quadratic structure can be considered as shown in Fig. 16.1b. There are two conditions in the relation between density and dispersity; one is that overlapping not occurs between two larger particles and the other is that overlapping not occurs between larger and smaller particles. These conditions can be expressed with σ as

$$\rho_{\mathrm{q}} < \frac{2(\sigma^2+1)}{(\sigma+1)^2} \quad \text{and} \tag{16.3}$$

$$\rho_{\mathrm{q}} < \sigma^2 + 1\,. \tag{16.4}$$

While the closest packing density of the hexagonal structure is a simply decreasing function, the closest density of the quadratic packing becomes larger with a larger dispersity, and will reach the maximum value of $\rho_{\mathrm{q}} = 4 - 2\sqrt{2}$ ($\simeq 1.17$) when $\sigma = \sqrt{2} - 1$ ($\simeq 0.414$) (see Fig. 16.1c). Note that the maximum density is larger than $2/\sqrt{3}$ ($\simeq 1.15$) which is the closest density of the monodisperse system. Therefore, it is possible that the quadratic solid structure is stable around $(\sigma, \rho) = (\sqrt{2}-1, 4-2\sqrt{2})$ in the dispersity-density parameter space.

16.3 Simulations

16.3.1 Bond-orientational Order Parameter

These two structures, hexagonal and quadratic, can be characterized by bond-orientational order (BOO) parameters [12]. The sixfold BOO parameter ϕ_6 is defined to be $\phi_6 = \langle \exp 6i\theta \rangle$, with the angle θ between a fixed axis and the bond connecting neighboring particles. The parameter ϕ_6 becomes 1 when the structure of the system is the perfect hexagonal, and becomes 0 when the structure is completely disordered. Therefore, ϕ_6 describes how close the system is to the perfect hexagonal structure. The fourfold parameter ϕ_4 is similarly defined to be $\phi_4 = \langle \exp(4i\theta) \rangle$.

16.3.2 Hexagonal Packing Configuration

We perform particle simulations and observe the nonequilibrium relaxation behavior of the BOO parameters. The starting configuration is set to be the perfect packing configuration, i.e., ($\phi_6(t=0) = 1$ or $\phi_4(t=0) = 1$). Periodic boundary conditions are taken for both directions of the simulation box. The particle number N is fixed at $23,288$ for the hexagonal and $10,000$ for the quadratic configurations, throughout the simulations. The time evolution of the system is performed by the event-driven molecular dynamics simulation. About 10^9 collisions are performed for each run and up to 512 independent samples are averaged for each density.

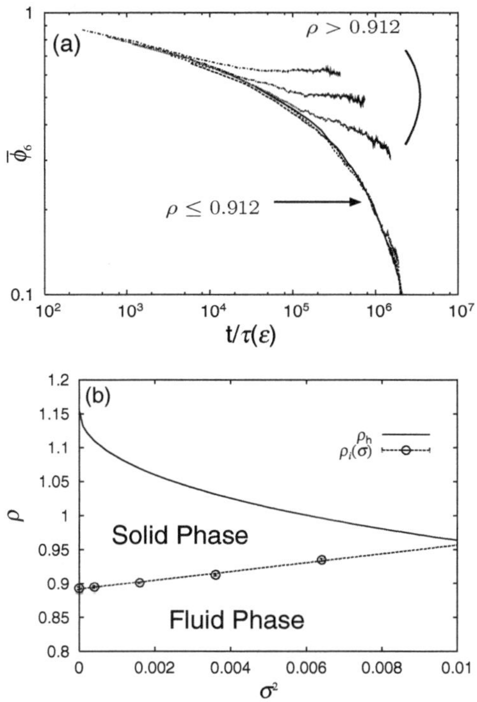

Fig. 16.2. (**a**) Scaled data for ϕ_6. The system of $\sigma = 0.06$. Decimal logarithm is taken for both *axes*. While the data of $\rho \leq 0.912$ are collapsed to the *single curve*, the data of $\rho \geq 0.914$ are not. (**b**) The shift of melting points between the isotropic and hexatic phases. The range of the hexatic phase is not shown since it is too narrow at the scale of this figure. The melting points are found to be almost linear to σ^2. The hexagonal solid phase can exist only in the region $\sigma < 0.1$

For the hexagonal packing configuration, the systems of $\sigma = 0$, 0.02, 0.04, 0.06 and 0.08 are studied. The melting transition of the monodisperse hard-disk system is predicted to be the KT transition [12] (see the review by Strandburg [13]), therefore, the dynamic KT scaling [8, 10, 11] is performed in order to determine melting points from the time evolution of ϕ_6. Near the critical point, a natural scaling form of the BOO parameter is expected to be

$$\phi_6(t, \varepsilon) = \tau(\varepsilon)^{-\lambda} \bar{\phi}_6(t/\tau(\varepsilon)) \quad (\varepsilon = \frac{\rho_i - \rho}{\rho_i}) , \tag{16.5}$$

with a correlation time τ, an exponent λ and a critical point ρ_i. Based on (16.5), the relaxation curves of $\tau^\lambda \phi_6$ plotted as a function of t/τ will collapse to a single curve with appropriately chosen λ and $\tau(\varepsilon)$.

In the KT transition, the correlation length diverges exponentially as $\xi \sim \exp(a'/\sqrt{\varepsilon})$ [14]. On the basis of the dynamical scaling hypothesis [15], the relation between ξ and τ is expected to be $\tau = \xi^z$ with a dynamic exponent

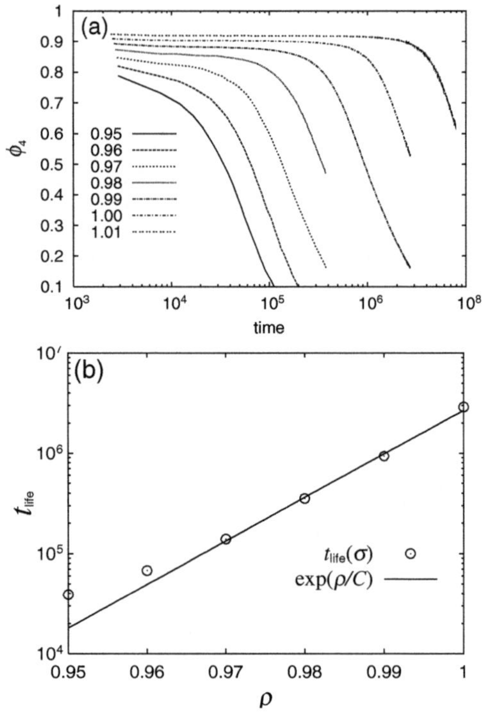

Fig. 16.3. (**a**) Relaxation behavior of ϕ_4 of the system with $\sigma = 0.4$ Decimal logarithm is taken for the *horizontal axis*. (**b**) Lifetime of the quadratic structure. Decimal logarithm is taken for the *vertical axis*. The lifetime is found to increase exponentially. The *solid line* is drawn for visual reference ($C = 0.01$)

z and the divergence behavior of the relaxation time τ is expected to be $\tau(\varepsilon) = b \exp(a/\sqrt{\varepsilon})$. The critical point can be determined by fitting this divergence behavior to $\tau(\varepsilon)$ obtained above.

The scaled result is shown in Fig. 16.2a. While the data of $\rho \leq 0.912$ are collapsed to the single curve, the data of $\rho \geq 0.914$ are not. The determined critical points for each dispersity are shown in Fig. 16.2b. The melting points are found to be almost proportional to the square of the dispersity. The hexagonal solid phase cannot exist in the region $\sigma > 0.1$ since the density of the melting transition will be higher than the highest density limit of the hexagonal packing configuration.

For the quadratic packing configuration, the systems which dispersities are $\sigma = 0.3$ to 0.5 are studied. For the cases of $\sigma = 0.3$ and 0.5, the quadratic structure was destroyed quickly even at the highest density. Therefore, if a quadratic solid exists, it is expected to be in the region of $0.3 < \sigma < 0.5$. The relaxation behavior of ϕ_4 at $\sigma = 0.4$ is shown in Fig. 16.3a. The lifetime $t_{\rm life}$ is defined by $\phi_4(t = t_{\rm life}) = 0.5$. The density dependence of the lifetime is shown in Fig. 16.3b. Although the transition behavior of the quadratic solid is not observed, the lifetime diverges exponentially.

16.4 Summary and Discussion

To summarize, hard-disk systems with equimolar bi-dispersity are studied. From the nonequilibrium relaxation behaviors of the bond-orientational order parameters, we find that (i) there is a critical dispersity at which the melting transition of the hexagonal solid vanishes and (ii) the quadratic structure is metastable in a certain region of the dispersity-density parameter space. These results suggest that the dispersity not only destroys orders but produces new structures under certain specific conditions.

In this work, only equimolar bi-disperse systems have been studied. The discussion of the relation between the closest density and the dispersity is difficult to apply for to polydisperse systems in general. The possible other structures are not considered here. These problems should be addressed in future studies.

Acknowledgements

This work was carried out on the SGI 2800 at the Supercomputer Center, Institute for Solid State Physics, University of Tokyo, and on the CP-PACS at the Center for Computational Physics, University of Tsukuba. We thank S. Miyashita for valuable discussions. This work was partly supported by the Nestlé Science Promotion Committee, the Grant-in-Aid for Scientific Research (C), No. 15607003, of Japan Society for the Promotion of Science, the Grant-in-Aid for Young Scientists (B), No. 14740229, of the Ministry of Education, Culture, Sports, Science and Technology of Japan, and the 21st COE program, "Frontiers of Computational Science", Nagoya University.

References

1. D.M. Mueth, H.M. Jaeger, S.R. Nagel, Phys. Rev. E **57**, 3164 (1998)
2. L. Prigozhin, H. Kalman, Phys. Rev. E **57**, 2073 (1998)
3. E. Dickinson, R. Parker, Chem. Phys. Lett. **79**, 578 (1981)
4. W. Vermölen, N. Ito, Phys. Rev. E **51**, 4325 (1995); see also Erratum Phys. Rev. E **54**, 3055 (1996)
5. M.R. Sadr-Lahijany, P. Ray, H.E. Stanley, Phys. Rev. Lett. **79**, 3206 (1997)
6. N. Ito, Physica A **192**, 604 (1993)
7. N. Ito, K. Hukushima, K. Ogawa, Y. Ozeki, J. Phys. Soc. Jpn., **69**, 1931 (2000)
8. Y. Ozeki, K. Ogawa, N. Ito, Phys. Rev. E **67**, 026702 (2003)
9. N. Ito, T. Matsuhisa, H. Kitatani, J. Phys. Soc. Jpn. **67**, 1188 (1998)

10. H. Watanabe, S. Yukawa, Y. Ozeki, N. Ito, Phys. Rev. E **66**, 041110 (2002); Note that the definition of the bond-orientational order was not correct. Therefore the critical point ρ_i without the dispersity was different from this study. We are preparing an erratum.
11. H. Watanabe, S. Yukawa, Y. Ozeki, N. Ito, Phys. Rev. E **69**, 045103(R) (2004)
12. B.I. Halperin, D.R. Nelson, Phys. Rev. Lett. **41**, 121 (1978); A.P. Young, Phys. Rev. B **19**, 1855 (1979)
13. K.J. Strandburg, Rev. Mod. Phys. **60**, 161 (1988)
14. J.M. Kosterlitz, D.J. Thouless, J. Phys. C **6**, 1181 (1973); J.M. Kosterlitz, J. Phys. C **7**, 1046 (1974)
15. B.I. Halperin, P.C. Hohenberg, Phys. Rev. **177**, 952 (1969)

Part IV

Methods

17

Asynchronously Parallelised Percolation on Distributed Machines

G. Pruessner[1] and N.R. Moloney[2]

[1] Physics Department, Virginia Polytechnic Institute and State University, Blacksburg, VA 24061-0435, USA
[2] ELTE, Institute for Theoretical Physics, Pázmány Péter sétány 1/a, 1117 Budapest, Hungary
Email: gunnar.pruessner@physics.org and moloney@general.elte.hu

Abstract. Based on the Hoshen-Kopelman algorithm for Monte-Carlo simulations of percolation, we propose a method to realise very large system sizes on distributed machines. It demands very little of hardware but nevertheless is very versatile and permits high precision measurements on very large lattices. The method has been used to calculate cluster size distributions as well as winding, spanning, and wrapping probabilities in two-dimensional percolation on various different topologies and aspect ratios.

17.1 Introduction

During its more than 60 year long history, percolation [1] has enjoyed the attention of physicists and mathematicians alike. With the appearence of conformal field theory (CFT) in statistical mechanics [2] it has experienced a renaissance. Cardy's seminal paper [3] on exact results for various spanning probabilities in two-dimensional percolation marks the first major breakthrough of CFT in percolation. At first glance this theoretical achievement makes numerical studies of 2D-percolation superfluous. Yet, there remain a number of open questions, such as non-universal quantities or, more importantly the relation between the conformal field theory studied and the corresponding lattice model [4]. Most notably, numerical studies have initially triggered the usage of CFT for percolation and produced the first clear signs of the existence of multiple spanning clusters in an appropriately defined thermodynamic limit [5–7].

Moreover, an analytical approach to higher dimensional systems is still lacking. In the following, an algorithm is presented that makes numerical simulations of very large systems possible, studying various very different properties at the same time and realising a wide range of aspect ratios. The technique is illustrated by a study of cluster size distributions and winding clusters on

Springer Proceedings in Physics, Volume 105
Computer Simulation Studies in Condensed-Matter Physics XVIII
Eds.: D.P. Landau, S.P. Lewis and H.-B. Schüttler

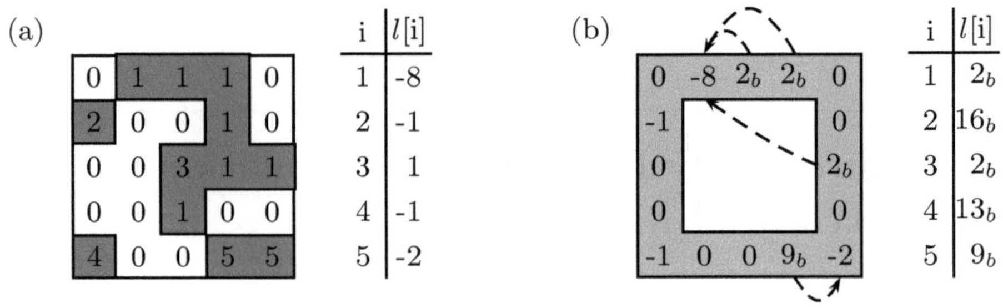

Fig. 17.1. (**a**) Example of a lattice configuration and its associated list of labels. Roots are negative numbers indicating the cluster size. (**b**) Border configuration after relabeling and path compression. From [10]

tori with various aspect ratios. It is based on the Hoshen-Kopelman algorithm (HKA) [8] which after almost 30 years remains the standard method for the identification of clusters. Strictly speaking, it is a data representation particularly suited for tracking clusters, which was recently exploited by Newman and Ziff [9] to study percolation very efficiently as a function of the occupation probability p.

Percolation is best defined as site-bond percolation, where bonds are active with probability p_b and sites occupied with probability p_s. In pure site percolation all bonds are active, $p_b = 1$, and correspondingly for bond percolation. A pair of neighbouring sites is connected provided that their bond is active and the sites occupied. If two sites are connected by a path along occupied sites and active bonds, they belong belong to the same cluster. Depending on the boundary conditions, these clusters have certain topological properties, such as their winding number.

What makes percolation crucially different from other problems in statistical mechanics is the easy availability of the "correct configuration distribution": Most Monte-Carlo simulations follow a particular path of configurations in the phase space generated by the underlying pseudo-dynamics, until the resulting trajectory (supposedly) produces a sequence of correlated but correctly, i.e. Boltzmann-distributed configurations. In contrast, an ensemble of percolation configurations is generated simply by occupying sites and activating bonds with a chosen probability. Without imposing the definition of clusters, there is no interaction between sites. Also, there are no correlations among different realisations and the equilibration time is exactly 0.

17.2 The Algorithm

In the following the algorithm is illustrated using the example of two-dimensional site percolation. In the HKA, the lattice is scanned row by row and a site is called active, if it possibly changes its connectivity, i.e. if it has

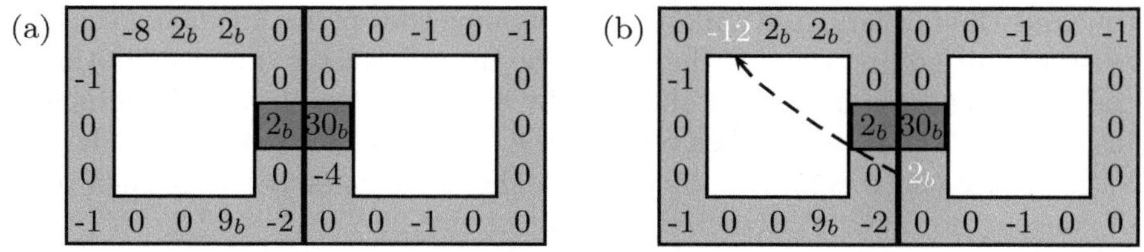

Fig. 17.2. Configuration of the border of two patches (**a**) before two clusters merge at the marked labels (**b**) after the merging, changed labels are shown in white. For uniqueness, the labels of the right patch are shifted by $4L-4$. From [10]

neighbours which have not been visited yet. To indicate the cluster it belongs to, every active site contains a pointer that refers to an entry in a list of labels, see Fig. 17.1a. These entries might point to other entries (the i in Fig. 17.1), a procedure which creates a tree on top of which the root contains all information (the $l[i]$ in Fig. 17.1) about the cluster, e.g. size, which borders it touches etc. Two sites then belong to the same cluster iff they point directly or indirectly to the same root.

Somewhat similar to Kadanoff's block spins [11], one can leave all $4L-4$ sites in the border of a freshly generated quadratic "patch" of length L active and visit them sequentially. The first site of previously unvisited clusters becomes the root of that cluster and replaces the entry from the list of labels by redirecting it. All information about the cluster is moved from the list of labels to that site. In Fig. 17.1b this is illustrated; labels "living" on the border carry an index b in Fig. 17.1. All other border sites belonging to the same cluster receive a label to point to that root site. The procedure is closely related to Nakanishi label recycling [12] and leads to a vector of length $4L-4$ containing all information about the clusters touching the boundaries of the patch.

Parallelising percolation, these patches are produced by a large number of "slave-nodes". The histograms of the clusters not touching the borders are stored locally, which requires only very little memory. The slaves send the resulting border configurations to a master node which compiles them into larger patches, see Fig. 17.2. Since percolation is effectively a non-interacting problem as described above, the border-configurations can arrive at the master nodes asynchronously. Whenever a sufficiently large number of them is available the master "glues" them together. The data presented below is based on a scheme where 22 slave nodes per master produce patches of size 1000×1000, 900 of which are glued together with different boundary conditions and aspect ratios. Correlations are reduced drastically by rotating, mirroring, and permuting the patches between different compilations.

As illustrated in Fig. 17.2, clusters on opposite sides of the glued borders merge by redirecting the root of the smaller cluster to point to the root of the larger one. If the histogram stored at the master contains all clusters touching the borders, it is updated by removing the entries for the two smaller clusters and adding one for the sum. The overall cluster size distribution of the resulting large patches is obtained by summing the slave and the master histograms. The large patches constructed can serve as the feed for a higher-level master which compiles the large patches to even larger ones. As a proof of concept, we have constructed a lattice of size $22.2 \cdot 10^6 \times 22.2 \cdot 10^6$, using this scheme.

17.3 Applications

As a first application, the overall cluster size distribution $n_{\mathrm{s,b}}(s;r)$ (the index indicates *s*ite or *b*ond percolation) has been studied, which is the total number of clusters of size s in a system with aspect ratio r, so that $\sum_s n_{\mathrm{s,b}}(s;r)s$ is the average number of occupied sites in the system. At the critical point one expects simple scaling,

$$n_{\mathrm{s,b}}(s;r) = a_{\mathrm{s,b}}(r)s^{-\tau}\mathcal{G}(s/s_0;r)\,, \tag{17.1}$$

where s_0 is a cutoff which is expected to scale like $b_{\mathrm{s,b}}(r)L^D$ with metric factor $a(r)_{\mathrm{s,b}}$ and $b(r)_{\mathrm{s,b}}$, critical exponents τ and D and universal scaling function $\mathcal{G}$. The metric factors are fixed against the scaling function by imposing two properties on it, such as its normalisation and its maximum. They are expected to differ for bond and site percolation.

The moment ratios

$$V_{m;\mathrm{s,b}} = \frac{\int ds n_{\mathrm{s,b}}(s;r)s^m}{\left(\int ds n_{\mathrm{s,b}}(s;r)s^2\right)^{m/2}} \tag{17.2}$$

are therefore expected to be non-universal, as they should differ by powers of the ratio

$$\frac{a_{\mathrm{s}}(r)/a_{\mathrm{b}}(r)}{(b_{\mathrm{s}}(r)/b_{\mathrm{b}}(r))^{\tau-1}}\,. \tag{17.3}$$

Most remarkably, this ratio is unity, suggesting that the ratio $a_{\mathrm{s,b}}(r)/b_{\mathrm{s,b}}^{\tau-1}(r)$, the concrete value of which depends on the conditions imposed on $\mathcal{G}$ and varies in r, is the same for bond and site percolation. In turn, this renders the ratios $V_{m;\mathrm{s,b}}$ (17.2) universal and in (17.1) one of the metric factors, say $a_{\mathrm{s,b}}$ could be replaced by $a_{\mathrm{s,b}}(r) = q(r)b_{\mathrm{s,b}}^{\tau-1}(r)$ with a function $q(r)$ depending (apart from the aspect ratio) only on the conditions imposed on $\mathcal{G}$.

Other applications are closely related to recent results from CFT and are concerned with, for example, the probability to find n distinct clusters that connect the top and the bottom of a lattice with cylindric boundary conditions [14], or to find n distinct clusters wrapping around such a cylinder. Similar measurements were done on a Möbius strip and a torus, where multiple distinct clusters can wind around horizontally and/or vertically (for details see [13]). To our knowledge, this is the first time winding clusters have been studied on the torus numerically; this is partly due to the rarity of some of these clusters, which we could observe with a frequency of only about 2 in 10^7 in patches of size 1000×1000 before they were subsequently used to build up the larger systems.

Details of this method and the results for two-dimensional site and bond percolations can be found in Refs. [10, 13, 14].

Acknowledgements

Most of the work presented has been performed at the Department of Mathematics and the Department of Physics at Imperial College London. The authors would like to thank K. Christensen for useful discussions as well as A. Thomas, M. Kaulke, O. Kilian, D. Moore and B. Maguire for their support. NRM is very grateful to the Beit Fellowship, and to the Zamkow family. GP gratefully acknowledges the support of the EPSRC, of the Alexander von Humboldt Foundation and of the NSF (DMR-0414122 and DMR-0308548).

References

1. P.J. Flory: J. Am. Chem. Soc. **63**, 3091 (1941)
2. R. Langlands, C. Pichet, P. Pouliot, Y. Saint-Aubin: J. Stat. Phys. **67**(3/4), 553 (1992)
3. J. Cardy: J. Phys. A: Math. Gen. **25**, L201 (1992); preprint hep-th/9111026
4. J.L. Cardy: Conformal invariance. In: *Phase Transitions and Critical Phenomena*. Vol. 11. Ed. by C. Domb, J.L. Lebowitz (Academic Press, London 1987)
5. C.-K. Hu, C.-Y. Lin: Phys. Rev. Lett. **77**(1), 8 (1996)
6. M. Aizenman: Nucl. Phys. B **485**, 551 (1997)
7. J. Cardy: J. Phys. A: Math. Gen. **31**, L105 (1998)
8. J. Hoshen, R. Kopelman: Phys. Rev. B **14**(8), 3438 (1976)
9. M.E.J. Newman, R.M. Ziff: Phys. Rev. Lett. **85**(19), 4104 (2000)
10. N.R. Moloney, G. Pruessner: Phys. Rev. E **67**, 037701-1 (2003); preprint cond-mat/0211240
11. L.P. Kadanoff: Physics **2**, 263 (1966)
12. K. Binder, D. Stauffer: Monte Carlo studies of "random" systems. In: *Applications of the Monte Carlo Method in Statistical Physics*. Vol. 36 of *Topics in Current Physics*, 2nd edn. Ed. by K. Binder (Springer, Berlin Heidelberg New York 1987)
13. G. Pruessner, N.R. Moloney: J. Stat. Phys. **115**(3/4), 839 (2004); preprint cond-mat/0310361
14. G. Pruessner, N.R. Moloney: J. Phys. A: Math. Gen. **36**(44), 11213 (2003); preprint cond-mat/0309126

18

Applications of AViz in Undergraduate Education

J. Adler and S. Rosen

Technion-IIT, Haifa, Israel, 32000

Abstract. Animated visualizations of physical systems can help undergraduate students understand and even enjoy their Physics classes. Preparing such visualizations provides interesting projects for senior undergraduate and graduate students, who learn basic techniques of computer simulation on systems that are relatively easy for them to understand.

18.1 Introduction

Many university teachers prefer research or discussions with their research students over undergraduate service-course teaching. Most undergraduate students likewise prefer playing computer games or hanging out with friends to attending lectures. However teaching undergraduates is survival for faculty, and for many of us its the way to spend the rest of the time on fun stuff like research, playing with computers etc. Studying physics is the meal ticket for undergraduates, and its the way they can get to the stage where they can spend the rest of their lives doing fun stuff like playing with computers etc. Thus since both enjoy playing with computers, if we make lectures more like computer games it is more fun for everyone and hopefully we increase the amount of time the students spend on physics.

Another common feature of many university departments is a final project for undergraduates. At the Technion this is either 3 or 4.5 hours a week for a 14 week semester. Computational projects are popular and in order to select a topic that the student can grasp within a reasonable timeframe, many projects relate to material from our three basic undergraduate physics classes, or from our undergraduate labs. The timeframe is tight because computational project students need time to learn the algorithm of simulation. An additional demand is the preparation of a website, usually in source-code html, to present the results. Students in our Computational Physics graduate course do similar projects, although many of these concern advanced topics they already understand. They use simulation techniques learnt in the course so the rela-

Springer Proceedings in Physics, Volume 105
Computer Simulation Studies in Condensed-Matter Physics XVIII
Eds.: D.P. Landau, S.P. Lewis and H.-B. Schüttler

tive credit and timeframe for the project is smaller. A useful by-product of these endeavours is a supply of teaching assistants who know how to develop websites for their classes.

Encouraging the students to prepare projects about topics from the three basic physics courses provides a souce of animations and programs that can liven up the lectures and turn dry material to something close to computer games.

18.2 Some Specifics

The first two elementary physics topics are usually mechanics and electricity and magnetism, which are quite amenable to lecture demonstration. However many topics from a third, "Modern Physics" course [1], such as Schroedinger equations and its applications to solid state and nuclear/particle physics cannot be demonstrated on the lecturer's bench. Fortunately many of these topics are ideal for simulation and for presentation on a website.

Our basic guidelines include the use of cheap legal software that is preferably public domain and at worst site-licensed at the Technion. To enliven the explanations visualization is important, and legal software often means LINUX nowadays so in practice most projects are either c/fortran with AViz (public domain) for atomistic visualization or matlab (site-licensed) for continuum systems (especially those of optics). Having applied these website rules since 1998 (before that an ftp server was used) it is nice to see that the 1998 pages still function OK as do many of the earlier routines.

Many condensed matter simulations are atomistic, and the early part of studying solid state physics involves learning about lattice geometry such as Bravais lattices and amorphous structures (as well as fivefold symmetries nowadays). Ball and stick lattice models, can of course show these, but 200 students see an on-screen animation more clearly. AViz [2,3] is a ball and stick visualization system designed for use in viewing animations of research simulations but it is equally useful for educational topics. AViz is public domain and creates ball and stick models from xyz data files via an interactive X11 interface. The graphics uses mesa/OpenGL, and there are options to change viewpoint, zoom, slice etc. The data files can be the result of a simulation or be calculated in real time as they are drawn, via the"watch" option. Bonds are drawn to aid the eye as needed at least for samples with less than a few hundred atoms.

18.3 An Example

One site [4] that has already been extensively used in courses at the Technion presents both Bravais and non-Bravais lattices as well as amino acids and

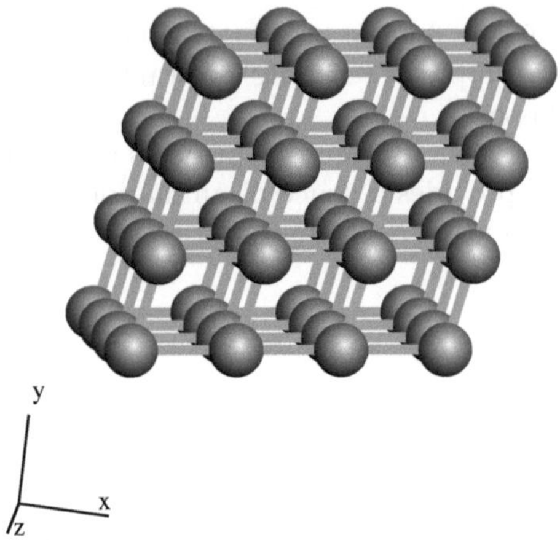

Fig. 18.1. Trigonal lattice

superconductors. As well as completed images in different sizes and animations of the lattice revolving to aid in three-dimensional sight, downloadable datafiles for most of the lattices, from which they can be reconstructed, are presented.

Let us take the trigonal lattice (Fig. 18.1) as an example. This image is built from a dataset of 64 atomistic positions whose first two lines give number of atoms and a comment line, followed by 64 more lines (6 shown here) with the type and the x, y and z coordinates of those 64 atoms.

64
c hg
hg 0 0 0
hg 0.47518459 0.47518459 2.92889471
hg 0.95036918 0.95036918 5.85778942
hg 1.42555377 1.42555377 8.78668413
hg 0.47518459 2.92889471 0.47518459
hg 0.95036918 3.4040793 3.4040793

To obtain this image write AViz and select this dataset [5] thru the "File" button. Select spheres, final quality and play with the bond length till you get only nearest neighbour bonds. More details are in an introductory article [6], and on the AViz website [2].

This site has become a standard for our Modern Physics classes. The Bravais lattice images are useful for the introduction to solid structures, the diamond structures for discussion of diamond and silicon and their energy gaps and the superconductor pictures to be shown alongside a demonstration of levitation. Similar images for other systems, including fullerenes, are at Ira Burshtein's site [7] which likewise includes xyz datafiles. One of these, C480, is shown in Fig. 18.2.

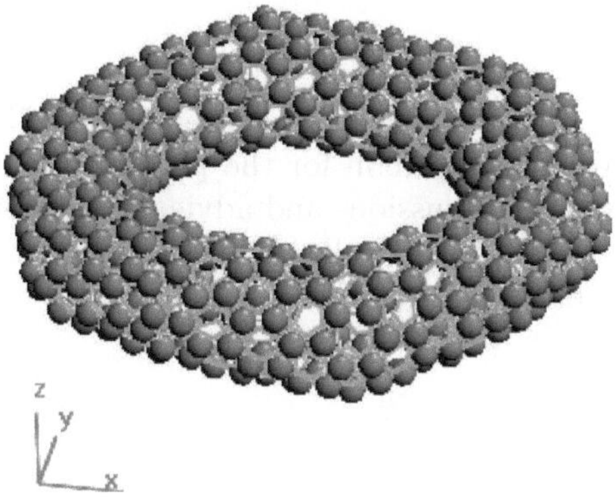

Fig. 18.2. A fullerence C480, data by Ira Burshtein

18.4 Other Topics, Other Graphics Systems

There is not time to list all the websites that have resulted from projects [8] here, but some AViz favourites include: "Bootstrap percolation" by Uri Lev [9], "Ising model" by Lior Metzger and "Internal structure of percolation clusters" by Eduardo Warszawski showing the famous red-green-blue bonds. All have downloadable programs to generate similar files with different random numbers or system sizes.

AViz is for atomistic/spin/polymer/pore visualizations. Not everything I want to show for Modern Physics is one of the above, and similar projects have been carried out over th elast 15 years with other graphics. A lot of the older 2D material (both continuum and discrete) is with PGPLOT (free for academic use and recommended in 2D) and much of the newer material is within MATLAB. MATLAB is not public domain, but we have a site-license and it is the overwhelming default in use at the Technion. Its graphics are quite bearable and it lends itself to interactive interfaces with ease.

Some of my favourite PGPLOT projects are: "Animated simulated annealing" with Amihai Silverman [10], which was our first interactive routine with pgplot, prepared to demonstrate the annealing of a sample of argon. A set of animated solutions of the Schrodinger equation have been prepared in sucessive projects by Yochai Ben Horin, Zaher Salman, Shai Goldberg and Yigal Ultchin. An interactive demostration of the Hoshen-Kopelman algorithm for percolation was prepared by Ido Braslavsky.

Some of my favourite MATLAB projects include: "Invasion percolation" by Dana Hoffmann, Polarization and Brewster's Angle by Nika Akopian and a Schrodinger equation interactive solver by Dany Regelman. Just this year three new routines designed for Modern Physics students were prepared by Yoav Hadas "Single Slit Diffraction", Katia Suhovoy, "X-ray diffraction as a way to study crystal structures" and Assaf Klein "Limit of resolution".

Acknowledgements

Shahar Rosen is now a graduste student at Ben Gurion University. We thank Geri Wagner and Adham Hashibon for the preparation of the current AViz version, and acknowledge discussions and advice from past and current Computational Physics students, especially the course and project students mentioned by name above.

References

1. for example R.A. Serway, J.W. Jewett Jr.: *Physics for scientists and engineers, with modern physics* (Australia: Thomson-Brooks/Cole 2004)
2. http://phycomp.technion.ac.il/~aviz
3. AViz AtomicVizualization, 2001 Computational Physics Group, Israel Institute of Technology Technion, 32000 Haifa Israel, Geri Wagner, Adham Hashibon
4. http://phycomp.technion.ac.il/~sshaharr
5. http://phycomp.technion.ac.il/~phr76ja/athens05/test.xyz
6. J. Adler: Comp. Sci. Eng. **5**, 61 (2003)
7. http://phycomp.technion.ac.il/~ira
8. http://phycomp.technion.ac.il/~comphy/technion_projects.html
9. J. Adler, U. Lev: Brazilian J. Phys. **33**, 641 (2003)
10. A. Silverman, J. Adler: Comp. Phys. **6**, 277 (1992)

19

Synchronous Sublattice Algorithm for Parallel Kinetic Monte Carlo Simulations

Y. Shim and J.G. Amar

Department of Physics and Astronomy, University of Toledo, Toledo, OH 43606

Abstract. A recently-developed efficient, semi-rigorous synchronous algorithm for parallel kinetic Monte Carlo simulations is presented. The accuracy and parallel efficiency are studied as a function of processor size, number of processors, and the ratio D/F of monomer hopping rate (D) to deposition rate (F) for a variety of simple models of epitaxial growth. Since only local communication is required, the algorithm scales, i.e. for a large number of processors the parallel efficiency is independent of the number of processors.

19.1 Introduction

While kinetic Monte Carlo (KMC) is an extremely efficient method [1–4] for simulating non-equilibrium processes, the standard KMC algorithm is inherently a serial algorithm, i.e. only one event can occur at each step. However, for some problems one needs to simulate larger length and time scales than can be simulated using a serial algorithm. Thus, it would be desirable to develop efficient parallel kinetic Monte Carlo algorithms so that many processors can be used simultaneously in order to carry out simulations over extended time and length scales.

Since in Metropolis Monte Carlo (MMC) [5] the event-time is independent of system configuration, rigorous and relatively efficient parallel MMC simulations may be carried out using an asynchronous "conservative" algorithm [6–10]. In contrast, in KMC the event time depends on system configuration. As a result, the problem of parallelizing the KMC algorithm is significantly more difficult [11]. We note that by mapping from KMC simulation to MMC simulation a rigorous algorithm for KMC simulations may be developed [10, 12]. However, as we have recently shown [12] the efficiency of such a hybrid algorithm is typically relatively low in simulations of thin-film growth due to the fact that the large range of event-rates leads to a high MMC rejection rate.

In order to try to improve the efficiency, recently we have studied [13, 14] the accuracy, parallel efficiency and scaling behavior of two different algo-

Springer Proceedings in Physics, Volume 105
Computer Simulation Studies in Condensed-Matter Physics XVIII
Eds.: D.P. Landau, S.P. Lewis and H.-B. Schüttler

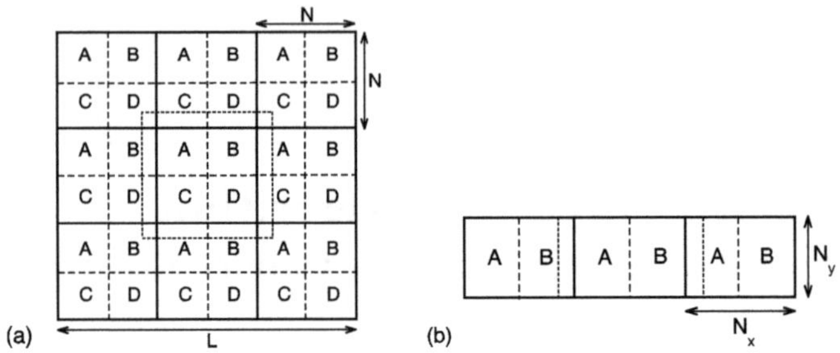

Fig. 19.1. Two possible methods of spatial and sublattice decomposition. (**a**) square sublattice decomposition (9 processors) and (**b**) strip sublattice decomposition (3 processors). *Solid lines* correspond to processor domains while *dashed lines* indicate sublattice decomposition. *Dotted lines* in (**a**) and (**b**) indicate "ghost-region" surrounding central processor

rithms for parallel kinetic Monte Carlo simulations of thin-film growth. The first algorithm corresponds to the rigorous synchronous relaxation (SR) algorithm suggested by Lubachevsky [15] in the context of dynamical Ising model simulations while the second algorithm is a semi-rigorous synchronous sublattice (SL) algorithm which we have recently developed [14]. Our results [13,14] indicate that both algorithms are relatively efficient and show reasonable scaling behavior. However, because the SL algorithm is both significantly faster and easier to implement than the SR algorithm, here we focus on the SL algorithm. We note that in some ways the SL algorithm is similar to that previously developed by Haider et al. [11]. However, in part because of the use of sublattices it has a number of significant advantages. In addition, it is significantly easier to implement and is also typically more efficient due to the reduced communications overhead.

19.2 Synchronous Sublattice Algorithm

In the synchronous sublattice (SL) algorithm, different parts of the system are assigned via spatial decomposition to different processors. However, in order to avoid conflicts between processors due to the synchronous nature of the algorithm, each processor's domain is further divided into different regions or sublattices (see Fig. 19.1). The SL algorithm may then be described as follows. At the beginning of a cycle each processor's local time is initialized to zero. One of the sublattices (e.g. A,B, C, or D in the square sublattice decomposition) is then randomly selected so that all processors operate on the same sublattice during that cycle. As in the usual serial KMC, each event is carried out with time increment $\Delta t_i = -\ln(r_i)/R_i$ where r_i is a uniform random number between 0 and 1 and R_i is the total KMC event rate for that processor's sublattice. Each processor then simultaneously and independently carries out KMC events in the selected sublattice until the time of the next event exceeds the time interval T. At the end of a cycle, each processor communicates any necessary changes (boundary events) with its neighboring processors, updates its event rates and then moves on to the next cycle using a new randomly chosen sublattice. We note that sublattice selection can be efficiently carried out without using communcation by seeding all processors with the same random number generator so that they all independently select the same sublattice for each cycle. By picking the cycle length T less than or equal to the average time for the fastest possible activated event we find that results are obtained which are indeed identical to those obtained in serial KMC except for very small sublattice sizes [14].

19.3 Results

A. Model. In order to test the performance and accuracy of our synchronous sublattice algorithm, we have studied a variety of irreversible and reversible solid-on-solid (SOS) models of epitaxial growth. These include an irreversible "fractal" growth model and an edge-and-corner diffusion (EC) model as well as more complex bond-counting models. In the "fractal" model [16], atoms (monomers) are deposited onto a square lattice with (per site) deposition rate F, diffuse to nearest-neighbor sites with hopping rate D and attach irreversibly to other monomers or clusters via a nearest-neighbor bond. The EC model is the same as the fractal model except that island relaxation is allowed, i.e. atoms which have formed a single nearest-neighbor bond with an island may diffuse along the edge of the island with diffusion rate $D_e = r_e D$ and around island-corners with rate $D_c = r_c D$. Both serial and parallel simulations were carried out using the 256-processor Itanium cluster at the Ohio Supercomputer Center (OSC) as well as the 3000-processor Alpha cluster at the Pittsburgh Supercomputer Center (PSC).

B. Comparison with Serial Simulations. Figure 19.2a shows a comparison of parallel and serial results for the r.m.s. surface roughness in the multilayer regime for the fractal model. As can be seen, there is essentially no difference between the serial and the parallel results.

Similar results are shown in Fig. 19.2b for the island-size distribution N_s (where N_s is the density of islands of size s) for the case of strip-decomposition. As can be seen, for reasonable processor sizes e.g. $N_x \geq 16$ there is again essentially no difference between the serial and parallel results. However, for extremely small processor sizes ($N_x = 8$ in Fig. 19.2b) there exists a small "finite-size" effect which leads to results which are slightly different from those obtained using the usual serial KMC algorithm. From simulations of a variety of models using both square-sublattice and strip-sublattice decomposition [14], we have found that there are no finite-size effects if the minimum processor size $N_{\min} > 2\,l_D$, where l_D corresponds to the diffusion length in submonolayer nucleation.

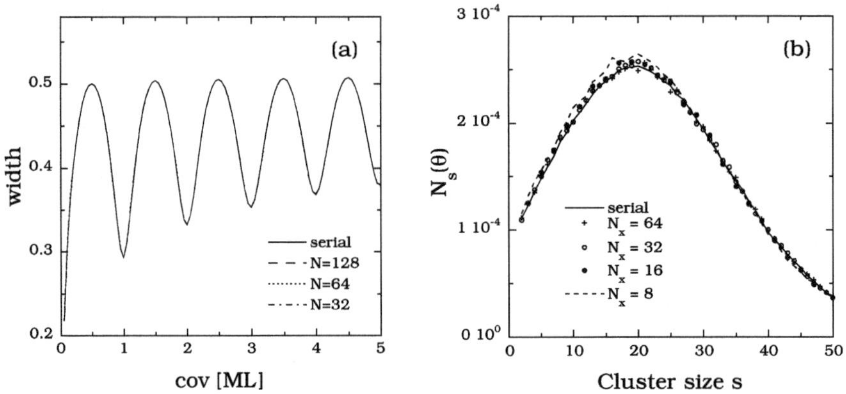

Fig. 19.2. Comparison between serial and parallel results for fractal model using both square and strip decomposition with $L = 256$ and $D/F = 10^5$. (**a**) Surface width as a function of coverage with $L = N \times \sqrt{N_p}$ and (**b**) island size distribution as a function of cluster size s with $N_x = 8, 16, 32$ and 64 and $N_y = 256$

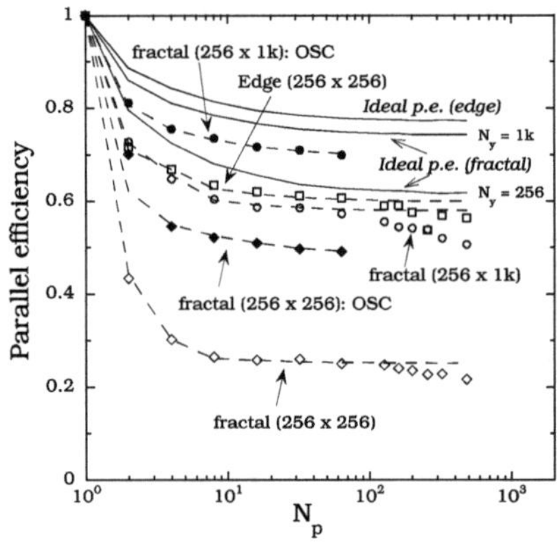

Fig. 19.3. Parallel efficiency (symbols) as function of number of processors N_p for fractal and EC models with $D/F = 10^5$, $\theta = 1\,\mathrm{ML}$, $r_e = 0.1$ and $r_c = 0$

C. Parallel Efficiency. The parallel efficiency PE may be obtained by calculating the ratio of the execution time t'_{1p} for an ordinary serial simulation of one processor's domain to the parallel execution time t_{N_p} of N_p domains using N_p processors, $PE = t'_{1p}/t_{N_p}$. Thus, the overall "performance factor" of the parallel simulation (e.g. boost in events/sec over a serial simulation) is given by the parallel efficiency multiplied by the number of processors N_p.

Figure 19.3 shows the parallel efficiency (symbols) for the fractal and EC models with $D/F = 10^5$ as a function of the number of processors N_p obtained from OSC and PSC. The ideal parallel efficiencies (solid lines) were obtained by assuming no communication overhead. As can be seen, in almost all cases the PE is of the order of 50% or more. In addition, since the SL algorithm only requires local communications, for a sufficiently large number of processors, the PE is essentially independent of the number of processors thus indicating perfect linear scaling.

19.4 Conclusion

We have developed and tested a synchronous sublattice algorithm for parallel kinetic Monte Carlo simulations. Since only local communication is required the algorithm scales, i.e. for a large number of processors the parallel efficiency is independent of the number of processors. For very small sublattice sizes, weak finite-size effects are observed which lead to deviations from the results obtained using a serial algorithm. However, since in many systems of interest the diffusion length is typically relatively small while significantly larger system sizes are needed to avoid finite system-size effects, the sublattice algorithm should provide a useful, efficient, and accurate method to carry out parallel KMC simulations of these systems. Due to its high parallel efficiency and relative simplicity compared to other algorithms such as the synchronous relaxation algorithm [13], we expect that the synchronous sublattice algorithm may be particularly useful for carrying out a variety of parallel non-equilibrium simulations.

References

1. A.B. Bortz, M.H. Kalos, J.L. Lebowitz: J. Comp. Phys. **17**, 10 (1975)
2. G.H. Gilmer: J. Crystal Growth **35**, 15 (1976)
3. A.F. Voter: Phys. Rev. B **34**, 6819 (1986)
4. J.L. Blue, I. Beichl, F. Sullivan: Phys. Rev. E **51**, R867 (1995)
5. N.C. Metropolis, A.W. Rosenbluth, M.N. Rosenbluth, A.H. Teller, E. Teller: J. Chem. Phys. **21**, 6 (1953)
6. K.M. Chandy, J. Misra: IEEE Trans. Software Eng. **5**, 440 (1979); J. Misra: ACM Comput. Surv. **18**, 39 (1986)
7. B.D. Lubachevsky: Complex Systems **1**, 1099 (1987); J. Comput. Phys. **75**, 103 (1988)
8. G. Korniss, Z. Toroczkai, M.A. Novotny, P.A. Rikvold: Phys. Rev. Lett. **84**, 1351 (2000)
9. G. Korniss, M.A. Novotny, H. Guclu, Z. Toroczkai, P.A. Rikvold: Science **299**, 677 (2003)
10. G. Korniss, M.A. Novotny, P.A. Rikvold: J. Comp. Phys. **153**, 488 (1999)
11. N. Haider, S.A. Khaddaj, M.R. Wilby, D.D. Vvedensky: Comput. Phys. **9**, 85 (1995)
12. Y. Shim, J.G. Amar: submitted to J. Comp. Phys.
13. Y. Shim, J.G. Amar: to appear in Physical Review B (http://lanl.arXiv.org/abs/condmat/0406540)
14. Y. Shim, J.G. Amar: to appear in Physical Review B (http://lanl.arXiv.org/abs/condmat/0406379)

15. B.D. Lubachevsky, A. Weiss, in *Proceeding of the 15th Workshop on Parallel and Distributed Simulation (PADS'01)* (IEEE, Piscataway, NJ, 2001)
16. J.G. Amar, F. Family, P.-M. Lam: Phys. Rev. B **50**, 8781 (1994)

20

Performance and Improvements of Flat-histogram Monte Carlo Simulations

Y. Wu[1], M. Köner[2], L. Colonna-Romano[3], S. Trebst[2], H. Gould[3], J. Machta[1], and M. Troyer[2]

[1] University of Massachusetts, Amherst MA 01003-3720, USA
[2] ETH, Zürich, CH-8093, Switzerland
[3] Clark University, Worcester, MA, 01610-1477, USA

Abstract. We study the performance of Monte Carlo simulations that samples a broad histogram in energy by determining the mean first passage time to span the entire energy space of d-dimensional Ising–Potts models. For the $d = 1, 2, 3$ Ising model, the mean first passage time τ of flat-histogram Monte Carlo methods with single-spin flip updates, such as the Wang–Landau algorithm or the multicanonical method, scales with the number of spins $N = L^d$ as $\tau \sim N^2 L^z$. The exponent z is found to be decrease as the dimensionality d is increased. In the mean field limit of infinite dimensions we find that z vanishes up to a logarithmic correction. We then demonstrate how the flat-histogram algorithms can be improved by two complementary approaches – cluster dynamics and ensemble optimization technique. Both approaches are found to improve the random walk in energy space so that $\tau \sim N^2$ up to logarithmic corrections for the $d = 1, 2$ Ising model.

20.1 Introduction

Monte Carlo (MC) methods sampling a canonical ensemble encounter difficulty in tunneling energy barriers. In order to deal with the problem, flat-histogram algorithms, which sample a uniform probability distribution in energy space, were invented. Recently, Wang and Landau [1, 2] introduced a flat-histogram algorithm that simulates a biased random walk in energy space, systematically estimate the density of states, $g(E)$. The algorithm has been applied to a wide range of systems.

One measure of the performance of the Wang-Landau algorithm is the mean first passage time τ, which we define to be the number of MC steps per spin for the system to go from the lowest energy to the highest energy. The number of possible energy values in the Ising model scales linearly with the number of spins $N = L^d$, where d is the spatial dimension and L is the linear size of the system. We expect $\tau \sim N^2$, if the Wang–Landau algorithm does an unbiased Markovian random walk. However, it was recently shown that τ scales as

Springer Proceedings in Physics, Volume 105
Computer Simulation Studies in Condensed-Matter Physics XVIII
Eds.: D.P. Landau, S.P. Lewis and H.-B. Schüttler

$$\tau \sim N^2 L^z \,, \tag{20.1}$$

where the exponent z is a measure of the deviation of the random walk from unbiased Markovian behavior, which is nonzero for Ising model [3].

In order to systematically study the performance of single-flip flat-histogram algorithm, we determine the mean first passage time for the ferromagnetic Ising model as a function of dimension d. We find that the exponent z is a decreasing function of d, and in the mean field limit of infinite dimensions z vanishes.

We then use two complementary approaches that improve the performance of flat-histogram methods. We show that the flat-histogram MC simulations can be speeded up by *changing spin dynamics*, that is, using cluster updates instead of local spin updates [4,5], or alternatively, by *changing the simulated statistical ensemble* to an optimal ensemble that minimize the mean first passage time [6].

20.2 Performance of Flat-histogram Algorithm with Local Updates

We measure the mean first passage time of single-spin flip flat-histogram algorithms for the $d = 1, 2, 3$ and mean field Ising model, and the result is listed below.

The errors are purely statistical. For $d = 1, 2$, we use exact density of states to do our measurement. For $d = 3$, we firstly estimate the density of states by the Wang–Landau algorithm and then make mean first passage time measurement.

In the mean field case the mean first passage time can be predicted by analytical methods described in [7]. We consider the infinite range Ising model with uniform interaction strength. In this system the energy E is simply related to the magnetization m, and it is convenient to express the density of states in terms of the latter. Let $t(m)$ be the mean first passage time starting from magnetization m, it then satisfies the following recursion relation

$$\begin{aligned} t(m) = {} & [t(m-\delta m) + \delta t]T_-(m) + [t(m+\delta m) + \delta t]T_+(m) \\ & + [t(m) + \delta t]T_0(m) \,. \end{aligned} \tag{20.2}$$

Table 20.1. z exponent of Ising Model

d	z
1	1.814 ± 0.014
2	0.743 ± 0.002
3	0.438 ± 0.010
∞ (mean field)	0 (logarithmic correction)

where $T_-(m) = T(m \to m - \delta m)$, $T_+(m) = T(m \to m + \delta m)$ are transition probabilities, and $T_0(m) = 1 - T_-(m) - T_+(m)$ is the waiting probability. This recursion relation is exactly solvable. For flat-histogram algorithms working in magnetization space, the mean fist passage time in the continuum limit is derived as

$$\tau = 2N^2 \int_{-1}^{1} \frac{1}{1 - |m| + 1/N} \, dm \sim N^2 \log N \; . \tag{20.3}$$

Therefore in mean field limit we prove rigorously that $z = 0$ up to a logarithmic correction, which is dominated by boundaries.

We are unable to extend the above method to Ising systems with nearest-neighbor couplings, because in these systems (20.2) no longer holds. We instead give a crude argument that, for the $d = 1$ Ising model, $z \leq 2$ comes from the domain wall diffusion process. Consider the two lowest states of a spin chain. The ground state has all spin aligned, and the first exited state has a single misaligned domain bounded by two domain walls. The distance between the domain walls is on average of order L, and therefore the typical number of MC steps required to make the transition from the first exited state to the ground state scales as L^2. If this diffusion time were to hold for all energies, the result would be $z = 2$. However, for higher energy states the domain wall diffusion time becomes smaller, and therefore $z = 2$ is an upper bound for the $d = 1$ Ising model.

20.3 Changing the Dynamics: Cluster Updates

In the previous section we have shown that the single-spin flip flat-histogram MC simulations suffer from a slowing down in Ising systems. The critical slowing down associated with single-spin flip algorithms in the canonical ensemble has been reduced by the introduction of cluster algorithms such as the Swendsen–Wang algorithm [8]. One can follow a similar approach and combine efficient cluster updates with flat-histogram simulations.

We will restrict our discussion to the Ising-Potts model. In the Swendsen–Wang algorithm the so-called spin-bond representation is employed, and the partition function of the Ising-Potts model is rewritten as

$$Z = \sum_{\sigma,\omega} p^{n_b - n(\omega)} (1-p)^{n(\omega)} \Delta(\sigma, \omega) \; , \tag{20.4}$$

where the summation is over spin configurations σ and bond configurations ω, $p = 1 - \mathrm{e}^{-\beta}$ is the occupying probability, $n(\omega)$ is the number of occupied bond, n_b is the number of bonds in the lattice, and $\Delta(\sigma, \omega) = 1$ if all occupied bonds are satisfied and zero otherwise.

It is natural to introduce a density of state $g(n)$ in terms of number of occupied bonds

$$g(n) = \sum_{\sigma,\omega} \delta(n - n(\omega)) \Delta(\sigma, \omega) \; . \tag{20.5}$$

The corresponding form of the partition function is

$$Z = \sum_n g(n) p^{n_b - n} (1-p)^n \,. \tag{20.6}$$

The only unknown in the partition function above is the density of states $g(n)$, and the Wang-Landau algorithm can be applied to estimate it. The resulting algorithm is called the multibondic algorithm [4, 5], which differs from the Swendsen–Wang algorithm only in that it uses a different acceptance probability of a bond move, $p_{acc} = \min[g(n)/g(n'), 1]$.

We measure the mean first passage time of the multibondic algorithm for $d = 1, 2$ Ising model, and we find $\tau \sim N^2 \log N$ in both cases [9].

20.4 Changing the Ensemble: Optimizing the Sample Histogram

An alternative to changing the dynamics from local to cluster update is to optimize the the simulated statistical ensemble and retain local spin updates. The ensemble $\omega(E)$ (or equivalently, the equilibrium distribution $n_\omega(E) \propto \omega(E) g(E)$) is optimized in the sense that it maximizes the round trip rate of the random walk. In [6] it is proved that the optimal equilibrium distribution is proportional to the inverse of the square root of the local diffusivity $D(E)$

$$n_\omega^{\mathrm{opt}}(E) \propto \frac{1}{\sqrt{D(E)}} \,, \tag{20.7}$$

and this optimal distribution can be achieved by feeding back $D(E)$. After the feedback the equilibrium distribution is no longer flat, but exhibit a peak at the region where the slowing down occurs, and thus the computing resources in the form of local updates are shifted towards the difficult region, which accelerates the algorithm.

We apply the optimal ensemble algorithm that uses single-spin flip Metropolis updates to the $d = 1$ Ising model. For $d = 1$ Ising model the algorithm has a reduced exponent $z = 0.972 \pm 0.002$, in comparison to single-spin flip flat-histogram simulations. The performance is further improved to $\tau \sim (N \log N)^2$ if we use the N-fold way updates [9]. For $d = 2$ Ising model $\tau \sim (N \log N)^2$ for both single spin-flip updates and N-fold way updates [6].

References

1. F. Wang and D.P.Landau: Phys. Rev. E **64**, 056101 (2001)
2. F. Wang and D.P.Landau: Phys. Rev. Lett. **86**, 2050 (2001)
3. P. Dayal., S. Trebst et al.: Phys. Rev. Lett. **92**, 097201 (2004)
4. W. Janke., S. Kappler: Phys. Rev. Lett. **74**, 212 (1995)

5. C. Yamaguchi., N. Kawashima: Phys. Rev. E **65**, 056710 (2002)
6. S. Trebst., D. Huse, M., Troyer: Phys. Rev. E **70**, 046701 (2004)
7. S. Redner: *A guide to First-Passge Processes*, (Cambrige University Press 2001)
8. R. H. Swendsen, J.-S. Wang, Phys. Rev. Lett **58**, 86 (1987)
9. Y. Wu, M.Köner et al: cont-mat/0412076

Part V

Biological and Soft Condensed Matter

21

Monte Carlo Study of the Isotropic–Nematic Interface in Suspensions of Spherocylinders

T. Schilling[1], R. Vink[1], and S. Wolfsheimer[1]

Johannes Gutenberg Universität Mainz, 55099 Mainz, Germany

Abstract. The isotropic to nematic transition in suspensions of anisotropic colloids is studied by means of grand canonical Monte Carlo simulation. From measurements of the grand canonical probability distribution of the particle density, the coexistence densities of the isotropic and the nematic phase are determined, as well as the interfacial tension.

21.1 Introduction

On change of density, suspensions of rod-like particles undergo a phase transition between an isotropic fluid phase, where the particle orientations are evenly distributed, and an anisotropic fluid phase (called "nematic" phase), where the particle orientations are on average aligned. Figure 21.1 shows a sketch of these phases.

In the 1940s, this phenomenon was explained by Lars Onsager in a theory based on infinitely elongated hard spherocylinders [1]. Onsager showed that the basic mechanism of the transition is the interplay between positional and orientational entropy. The size of the excluded volume – i.e. the volume around one particle, which another particle cannot enter, because it would produce an overlap – depends on the angle between the two particles' axis. If the

Fig. 21.1. Sketches of the isotropic phase (*left*) and nematic phase (*right*): In the isotropic phase, particle positions and orientations are disordered. In the nematic phase, particle orientations are aligned

Springer Proceedings in Physics, Volume 105
Computer Simulation Studies in Condensed-Matter Physics XVIII
Eds.: D.P. Landau, S.P. Lewis and H.-B. Schüttler

particles lie parallel, the excluded volume is minimized. Hence particles which are aligned, gain accessible volume and therefore positional entropy, but they loose orientational entropy. At a certain density the balance between the two contributions flips and the system changes from the isotropic to the nematic phase.

Onsager theory has been remarkably successful at describing the isotropic to nematic (IN) transition, and still serves as the basis for many theoretical investigations of the properties of liquid crystals. Over the last twenty years, for instance, several groups have investigated the properties of the interface between the two phases using Onsager–type density functional approaches [2–8].

The IN interface is an interesting model system, because both phases are almost incompressible and their densities are similar. Therefore pressure and density are not important parameters. The only parameters which determine the properties of the interface are the particle aspect ratio and the orientation of the director (i.e. the axis of average orientation of the particles in the nematic phase) with respect to the plane of the interface.

An important finding of the theoretical studies cited above is that the interfacial tension γ_{IN} of the IN interface is minimized when the director lies in the plane of the interface. In this case γ_{IN} is predicted to be very low, but the precise value varies considerably between different authors [9,10]. Theoretical estimates for γ_{IN} typically range from 0.156 [7] to 0.34 [3], in units of $k_B T/LD$, with L the rod length, D the rod diameter, T the temperature, and k_B the Boltzmann constant.

To test the accuracy of the theoretical estimates of γ_{IN}, one might wish to make a direct comparison to experimental data. Unfortunately, this is not straightforward. The models used in theoretical treatments of the IN interface are typically rather simplistic, usually based on a hard or short-ranged pair potential in a system of monodisperse spherocylinders. Using these models, it is not reasonable to expect quantitative agreement with experiments, because the interactions in the experimental system will be much more complex. For example, polydispersity may be an important factor, and it is not clear to what extent long-range interactions play a role. Also many experimental systems display chirality. And even the experimental determination of the rod dimensions L and D, required if a comparison to theory is to be made, presents complications [10].

In order to validate the assumptions made by the various approaches, it is nevertheless important to test the accuracy of the theoretical predictions. To this end, computer simulations are ideal, because they, in principle, probe the phase behavior of the model system without resorting to approximations. In recent years, several groups have investigated the IN transition by means of simulations [11–18]. However, the interfacial tension γ_{IN} was not measured in these studies.

To obtain γ_{IN} in simulations rather elaborate simulation techniques are required. One possibility is to measure the anisotropy of the pressure tensor.

The interfacial tension is obtained from the difference between the normal and the transversal pressure tensor components:

$$\gamma_{\mathrm{IN}} = \int_{-\infty}^{\infty} P_N(z) - P_T(z)\,\mathrm{d}z\,, \tag{21.1}$$

where the interface lies in the xy-plane. In [18], this method is applied to suspensions of ellipsoids with axial ratio $\kappa = A/B = 15$, where A is the length of the symmetry axis, and B that of the transverse axis. The measured interfacial tension is $0.006 \pm 0.005\,k_BT/B^2 \approx 0.09\,k_BT/AB$ if a hard interaction potential is used, and $0.011 \pm 0.004\,k_BT/B^2 \approx 0.165\,k_BT/AB$ for a soft potential. The anisotropy of the pressure tensor is very small, and thus difficult to measure accurately. Therefore the error bars of these results are large.

Another approach is via the capillary wave spectrum. The basic idea is the following: the interface will fluctuate for entropic reasons. As enlargement of the interfacial area costs energy, the spectrum of the fluctutions is related to γ_{IN}. If the interface is described by a function $h(x, y)$, then one can show that

$$\langle |h(\boldsymbol{q})|^2 \rangle = \frac{k_BT}{\gamma_{\mathrm{IN}} q^2}\,, \tag{21.2}$$

where $h(\boldsymbol{q})$ is the Fourier transformed of $h(x, y)$ [19]. In [19] this approach was applied to soft ellipsoids with $\kappa = 15$. $\gamma_{\mathrm{IN}} = 0.016 \pm 0.002\,k_BT/B^2 \approx 0.24\,k_BT/AB$ is reported. However, capillary wave theory is only valid in the long wavelength limit. Therefore very large system sizes are required. Moreover, if periodic boundary conditions are used, two interfaces will be present in the simulation box. Since γ_{IN} is very small, large capillary fluctuations can occur, and one needs to be aware of interactions between the two interfaces. Therefore this method requires very large system sizes.

Clearly, in order to obtain γ_{IN} more accurately, much more computer power or different simulation techniques are required. In this article we present a method, which allows to reduce statistical errors considerably and therefore makes an analysis of the finite-size effects possible. Recent advances in grand canonical sampling methods [20, 21] have enabled accurate measurements of the interfacial tension in simple fluids [22, 23] and colloid–polymer suspensions [24, 25]. The aim of this paper is to apply these techniques to the IN transition in a system of soft spherocylinders, and to extract the corresponding phase diagram and the interfacial tension. Simulations in the grand canonical ensemble offer a number of advantages over the more conventional methods discussed previously. More precisely, in grand canonical simulations, both the coexistence properties can be probed, as well as the interfacial properties – where as the methods described above require an independent estimate of the coexistence densities. Additionally, finite–size scaling methods are available which can be used to extrapolate simulation data to the thermodynamic limit [26–29].

This article is structured as follows: First, we introduce the soft spherocylinder model used in this work. Next, we describe the grand canonical Monte

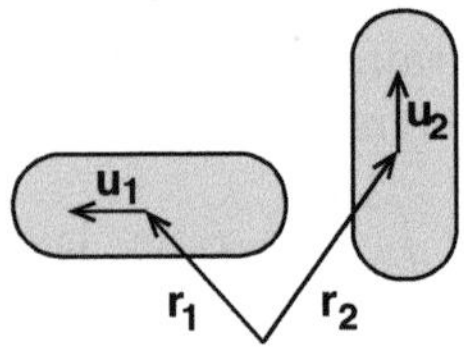

Fig. 21.2. Definition of the quantities used in (21.3)

Carlo method, and explain how the coexistence properties, and the interfacial tension are obtained. Finally, we present our results.

21.2 Model

For numerical reasons, which will be explained in Sect. 21.3, we do not model the particles as hard rods, but as repulsive soft rods. Two spherocylinders of elongation L and diameter D interact via a pair potential of the form

$$V(\boldsymbol{r}_1, \boldsymbol{r}_2, \boldsymbol{u}_1, \boldsymbol{u}_2) = \begin{cases} \varepsilon & : \quad r < D \\ 0 & : \quad \text{otherwise} \end{cases} , \tag{21.3}$$

where $\boldsymbol{r}_1$, $\boldsymbol{r}_2$, $\boldsymbol{u}_1$ and $\boldsymbol{u}_2$ are definied in Fig. 21.2 and r is the distance between the particles' axis. The total energy is thus proportional to the number of overlaps in the system. In this article, the rod diameter D is taken as unit of length, and k_BT as unit of energy. The strength of the potential is set to $\varepsilon = 2\,k_BT$. Note that in the limit $\varepsilon \to \infty$, this model approaches a system of infinitely hard rods. The effect of the particle softness is only a shift in the coexistence densities to higher values. The transition mechanism remains the same.

To study the IN transition, we use the density and the rod alignment as order parameters. Both the isotropic and the nematic phase are fluid phases, in the sense that long-range positional order of the centers of mass is absent. Orientational order is measured by the S_2 order parameter, defined as the maximum eigenvalue of the thermal average of the orientational tensor $\underline{Q}$:

$$Q_{\alpha\beta} = \frac{1}{2N} \sum_{i=1}^{N} (3u_{i\alpha}u_{i\beta} - \delta_{\alpha\beta}) \ . \tag{21.4}$$

Here, $u_{i\alpha}$ is the α component ($\alpha = x, y, z$) of the orientation vector $\boldsymbol{u}_i$ of rod i (normalized to unity), and $\delta_{\alpha\beta}$ is the Kronecker delta. In the case of orientational order S_2 assumes a value close to one, while in the disordered isotropic phase, S_2 is close to zero.

Since the density of the nematic phase is slightly higher than that of the isotropic phase, we may also use the particle number density $\rho = N/V$ to distinguish between the phases, with N the number of rods in the system, and

V the volume of the simulation box. Following convention, we also introduce the reduced density $\rho^\star = \rho/\rho_{cp}$, with $\rho_{cp} = 2/(\sqrt{2}+(L/D)\sqrt{3})$ the density of regular close packing of hard spherocylinders.

21.3 Simulation Method

The simulations are performed in the grand canonical ensemble. In this ensemble, the volume V, the temperature T, and the chemical potential μ of the rods are fixed, while the number of rods N inside the simulation box fluctuates. Insertion and removal of rods are attempted with equal probability, and accepted with the standard grand canonical Metropolis rules, given by

$$A(N \to N+1) = \min\left[1, \frac{V}{N+1}\exp\left(-\beta\Delta E + \beta\mu\right)\right] \tag{21.5}$$

and

$$A(N \to N-1) = \min\left[1, \frac{N}{V}\exp\left(-\beta\Delta E - \beta\mu\right)\right], \tag{21.6}$$

where ΔE is the energy difference between initial and final state, and $\beta = 1/k_BT$ [27,30]. Here it becomes evident, why this method is difficult to apply to a system of hard objects – insertion moves will become extremely unlikely, if overlaps are forbidden. Therefore we introduced a finite energy cost instead.

The simulations are performed in a three dimensional box of size $L_x \times L_y \times L_z$ using periodic boundary conditions in all directions. In this work, we fix $L_x = L_y$, but we allow for elongation $L_z \geq L_x$. Moreover, to avoid double interactions between rods through the periodic boundaries, we set $L_x > 2L$.

During the simulations, we measure the probability distribution $P(N)$, defined as the probability of observing a system containing N rods. The shape of the distribution will depend on the following parameters:

- the aspect ratio L/D,
- the chemical potential μ,
- the box dimensions L_x and L_z, because there will be finite-size effects,

and not on T, because it just sets the energy scale. At phase coexistence, the distribution $P(N)$ becomes bimodal, with two peaks of equal area, one located at small values of N corresponding to the isotropic phase, and one located at high values of N corresponding to the nematic phase. A typical coexistence distribution is shown in Fig. 21.3, where the logarithm of $P(N)$ is plotted.

In order to find the chemical potential of coexistence, we use the equal area rule [31]. Coexistence is defined as the situation in which the areas under the peaks are equal:

$$\int_0^{\langle N\rangle} P(N)\mathrm{d}N = \int_{\langle N\rangle}^{\infty} P(N)\mathrm{d}N\,, \tag{21.7}$$

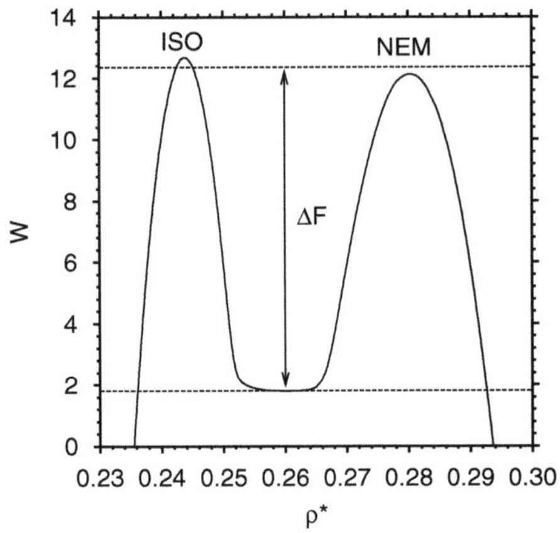

Fig. 21.3. Coexistence distribution $W = k_B T \ln P(N)$ of the isotropic to nematic transition in a system of soft rods with $\varepsilon = 2$ and $L/D = 15$. The low density peak corresponds to the isotropic phase (ISO), the high density peak to the nematic phase (NEM), and the barrier ΔF to the free energy difference between the two phases (ΔF is given by the average peak height as measured from the minimum in between the peaks). The above distribution was obtained using box dimensions $L_x = 2.1L$ and $L_z = 8.4L$

where $\langle N \rangle$ is the average of the full distribution

$$\langle N \rangle = \int_0^\infty N P(N) \mathrm{d}N \,, \tag{21.8}$$

and we assume that $P(N)$ has been normalized to unity $\int_0^\infty P(N)\mathrm{d}N = 1$. The coexistence density of the isotropic phase follows trivially from the average of $P(N)$ in first peak

$$\rho_{\text{iso}} = (2/V) \int_0^{\langle N \rangle} N P(N) \mathrm{d}N \,, \tag{21.9}$$

and similarly for the nematic phase

$$\rho_{\text{nem}} = (2/V) \int_{\langle N \rangle}^\infty N P(N) \mathrm{d}N \,, \tag{21.10}$$

where the factors of two are a consequence of the normalization of $P(N)$.

The interfacial tension γ_{IN} is extracted from the logarithm of the probability distribution $W \equiv k_B T \ln P(N)$. Since $-W$ corresponds to the free energy of the system, the average height ΔF of the peaks in W, measured with respect to the minimum in between the peaks, equals the free energy barrier separating the isotropic from the nematic phase. A snapshot (Fig. 21.4) shows, which configurations contribute to $P(N)$ in the region where the overall density of the system is between the peaks $\rho_{\text{iso}} \ll \rho \ll \rho_{\text{nem}}$. We have a slab

Fig. 21.4. Snapshot of a system of soft spherocylinders at IN coexistence. The spherocylinders are shaded according to their orientation. On the left side of the *dashed line* the system is isotropic, on the right side it is nematic. The second interface coincides with the boundaries of the box in the elongated direction

geometry, with one isotropic region, and one nematic region, separated by an interface (because of the periodic boundary conditions, there are actually two interfaces). Note that the director of the nematic phase lies in the plane of the interfaces. This was the typical case for the snapshots studied by us, and is consistent with the theoretical prediction that in-plane alignment yields the lowest free energy. From this snapshot it becomes evident why the distribution has a flat region. If the two interfaces are sufficiently separated they do not interact. Then it is possible to change the overall system density simply by growing one phase and shrinking the other without producing any free energy.

The barrier ΔF in Fig. 21.3 thus corresponds to the free energy cost of creating two interfaces in the system. In this work, where the box dimensions are chosen such that $L_x = L_y$ and $L_z \geq L_x$, the interfaces will be oriented perpendicular to the elongated direction, since this minimizes the interfacial area, and hence the free energy of the system. The total interfacial area in the system thus equals $2L_x^2$. As the interfacial tension is simply the excess free energy per unit area, we may write

$$\gamma_{\mathrm{IN}}(L_x) = \Delta F/(2L_x^2)\,, \tag{21.11}$$

with $\gamma_{\mathrm{IN}}(L_x)$ the interfacial tension in a finite simulation box with lateral dimension L_x [26].

To obtain the interfacial tension in the thermodynamic limit, one can perform a finite size scaling analysis [26] to estimate $\lim_{L_x \to \infty} \gamma_{\mathrm{IN}}(L_x)$. Alternatively, away from any critical point, the most dominant finite size effects will likely stem from interactions between the two interfaces. In this case, it is feasible to use an elongated simulation box with $L_z \gg L_x$, such as in Fig. 21.4. (This supresses interactions between the interfaces, simply because their distance is larger.) In this article we will focus on the second aspect. A finite size analysis of our results can be found in [32].

If the free energy barrier ΔF is large, transitions between the isotropic and the nematic phase become less likely, and the simulation will spend most of the time in only one of the two phases. A crucial ingredient in our simulation is therefore the use of a biased sampling technique. We use successive umbrella sampling [21] to enable accurate sampling in regions where

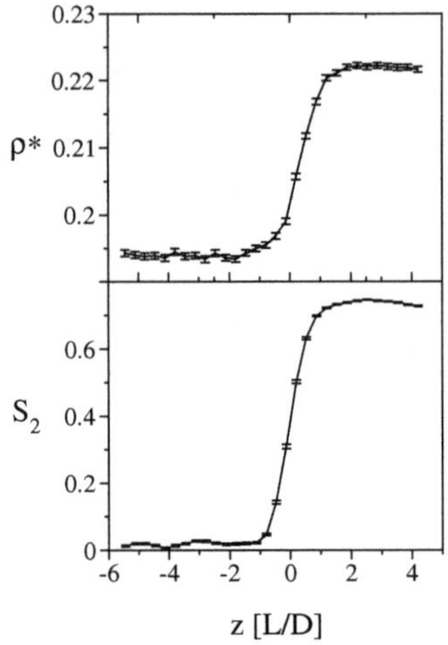

Fig. 21.5. Profiles of density and nematic order perpendicular to the interface for $L/D = 15$. In agreement to Onsager theory, the profiles are shifted by roughly $0.3\,L/D$ with respect to one another

$P(N)$, due to the free energy barrier separating the phases, is very small. Note also that phase coexistence is only observed if the chemical potential μ is set equal to its coexistence value. This value is in general not known at the start of the simulation, but it may easily be obtained by using the equation $P(N|\mu_1) = P(N|\mu_0)e^{\beta(\mu_1-\mu_0)N}$, with $P(N|\mu_\alpha)$ the probability distribution $P(N)$ at chemical potential μ_α. In the simulations, we typically set the chemical potential to zero and use successive umbrella sampling to obtain the corresponding probability distribution. We then use the above equation to obtain the desired coexistence distribution, in which the area under both peaks is equal.

21.4 Results

21.4.1 Profiles

First we estimated the minimum box size necessary to contain two independent interfaces. We ran an NVT Monte Carlo simulation of hard spherocylinders at coexistence in a very elongated box ($L_z = 10\,L$) and measured the interfacial width. Figure 21.5 shows the density (top) and the nematic order parameter (bottom) perpendicular to the interface. We do not see any trace of non-monotonous behaviour. In agreement with Onsager theory, the interfaces are shifted with respect to one another. Moving from the nematic to the isotropic region, first ρ drops and then S_2. Measured from the point where the density is $0.9\rho_{\text{iso}}$ to the point, where it is $0.9\rho_{\text{nem}}$ the interface is roughly

4 L wide. This means that very large simulation boxes will be necessary to properly decouple two interfaces.

21.4.2 Phase Diagram

First we used our grand canonical Monte Carlo scheme to determine the IN phase diagram of the soft spherocylinder system of (21.3) using $\varepsilon = 2$. For several rod elongations L/D, we measured the distribution $P(N)$, from which $\rho_{\rm iso}$ and $\rho_{\rm nem}$ were obtained. The system size used in these simulations is typically $L_x = L_y = 2.1L$ and $L_z = 4.2L$. In Fig. 21.6, we plot the reduced density of the isotropic and the nematic phase as function of L/D. We observe that the phase diagram is qualitatively similar to that of hard spherocylinders [12]. The quantitative difference being that, for soft rods, the IN transition is shifted towards higher density. The inset of Fig. 21.6 shows the concentration variable $c = \pi D L^2 \rho/4$ as a function of D/L. For hard spherocylinders, Onsager theory predicts that $c_{\rm ISO} = 3.29$ and $c_{\rm NEM} = 4.19$ in the limit of infinite rod length, or equivalently $D/L \to 0$. In case of the soft potential of (21.3), these values must be multiplied by $(1 - e^{-\beta\varepsilon})^{-1} \approx 1.16$ for $\varepsilon = 2$. In the inset of Fig. 21.6, the corresponding limits are marked with arrows. As in [12], we observe that the simulation data for the isotropic phase smoothly approach the Onsager limit, while the nematic branch of the binodal seems to overshoot the Onsager limit. This we attribute to equilibration problems. To simulate the IN transition in the limit $D/L \to 0$, large system sizes are required, and it becomes increasingly difficult to obtain accurate results. To quantify the uncertainty in our measurements, additional independent simulations for rod elongation $L/D = 25$, 30, and 35 were performed. The corresponding data are also shown in Fig. 21.6. For $L/D \geq 30$, we observe significant scatter, while for $L/D \leq 25$, the uncertainty is typically smaller than the symbol size used in the plots.

21.4.3 Interfacial Tension

Next, the interfacial tension $\gamma_{\rm IN}$ is determined for $L/D = 10$ and $L/D = 15$. Unfortunately, the system size used to compute the phase diagram in the previous section, was insufficient to accurately extract the interfacial tension because no flat region between the peaks in $P(N)$ could be distinguished. To properly extract the interfacial tension, much larger systems turned out to be required. In this case, care must be taken in the sampling procedure. Many sampling schemes, especially the ones that are easy to implement such as successive umbrella sampling, put a bias on the density only. Such schemes tend to "get stuck" in meta–stable droplet states when the system size becomes large [24]. As a result, one may have difficulty reaching the state with two parallel interfaces, in which case (21.11) cannot be used.

Therefore, for large systems, one must carefully check the validity of the simulation results. We performed a number of additional grand canonical simulations using a biased Hamiltonian of the form $\mathcal{H} = \mathcal{H}_{\prime} + \mathcal{W}$, with $\mathcal{H}_{\prime}$ the

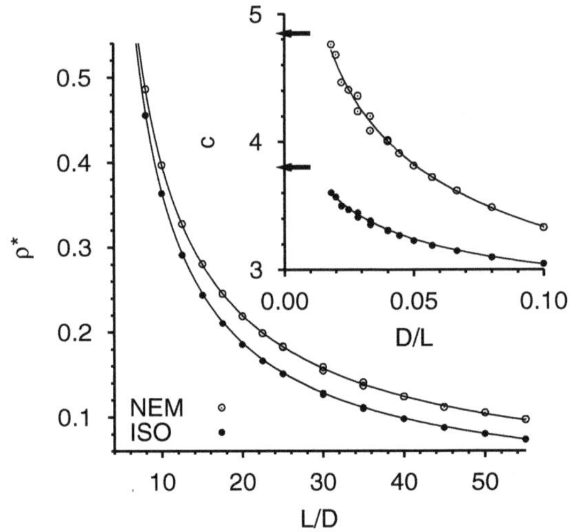

Fig. 21.6. Soft spherocylinder phase diagram of the IN transition using $\varepsilon = 2$. Shown is the reduced density $\rho^\star$ of the isotropic phase (*closed circles*) and of the nematic phase (*open circles*) as function of L/D. The inset shows the concentration variable c as function of D/L for both the isotropic and the nematic phase. The lower and upper arrow in the inset mark the Onsager limit $D/L \to 0$ for the isotropic and the nematic phase, respectively. The *lines connecting the points* serve as a guide to the eye

Hamiltonian of the real system defined by (21.3) and $W = -k_BT \ln P(N)$. If the measured $P(N)$ is indeed the equilibrium coexistence distribution of the real system, a simulation using the biased Hamiltonian should visit the isotropic and the nematic phase equally often on average [24,33]. This is illustrated in the top frame of Fig. 21.7, which shows the S_2 order parameter as a function of the elapsed simulation time during one such biased simulation. Indeed, we observe frequent transitions between the isotropic ($S_2 \sim 0$) and the nematic phase ($S_2 \sim 1$). Also shown in Fig. 21.7 is the corresponding time series of the reduced density. In case that a perfect estimate for $P(N)$ could be provided, the measured distribution in the biased simulation will become flat in the limit of long simulation time. The deviation from a flat distribution can be used to estimate the error in $P(N)$, or alternatively, to construct a better estimate for $P(N)$. The latter approach was adopted by us. First, successive umbrella sampling is used to obtain an initial estimate for $P(N)$. This estimate is then used as input for a number of biased simulations using the modified Hamiltonian, and improved iteratively each time.

To obtain the interfacial tension, the most straightforward approach is to fix the lateral box dimensions at $L_x = L_y$, and to increase the elongated dimension $L_z \gg L_x$ until a flat region between the peaks in the distribution $P(N)$ appears. For soft spherocylinders of elongation $L/D = 10$, the

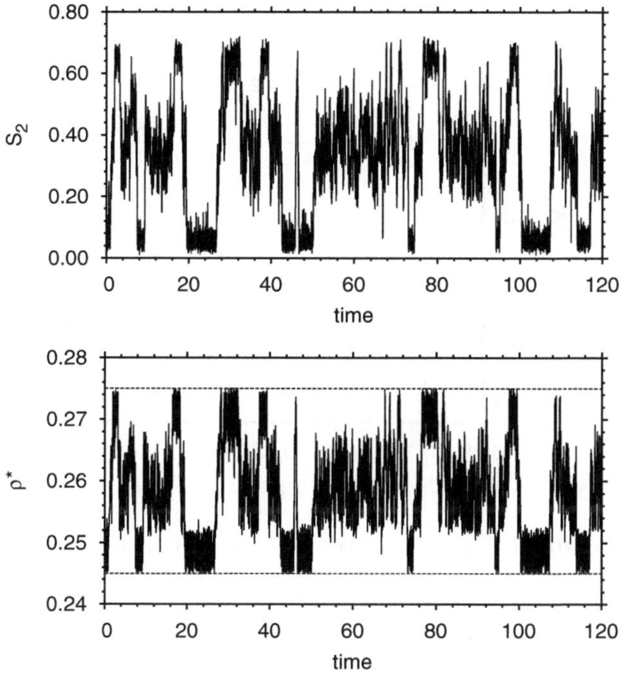

Fig. 21.7. Monte Carlo time series of a biased grand canonical simulation. The *top frame* shows the S_2 order parameter as a function of the invested CPU time, the *lower frame* the reduced density, with CPU time expressed in hours on a 2.6 GHz Pentium. During the simulation, the reduced density was confined to the interval $0.245 < \rho^\star < 0.275$, as indicated by the *horizontal lines* in the lower figure. The data were obtained using $L/D = 15$, $\varepsilon = 2$, $L_x = 2.1L$ and $L_z = 8.4L$, which are the same parameters as used in Fig. 21.3

results of this procedure are shown in Fig. 21.8. Indeed, we observe that the region between the peaks becomes flatter as the elongation of the simulation box is increased. Unfortunately, even for the largest system that we could handle, the region between the peaks still displays some curvature. In other words, the interfaces are still interacting, indicating that even more extreme box elongations are required. Ignoring this effect, and applying (21.11) to the largest system of Fig. 21.8, we obtain for the interfacial tension $\gamma_{\mathrm{IN}} = 0.0022\, k_B T/D^2$. For rod elongation $L/D = 15$, the distribution of the largest system that we could handle is shown in Fig. 21.3. The height of the barrier reads $\Delta F = 10.59\, k_B T$, and the corresponding interfacial tension $\gamma_{\mathrm{IN}} = 0.0053\, k_B T/D^2$.

The advantage of the present simulation approach is that the statistical errors are small, and that finite size effects are clearly visible as a result. In contrast, if the pressure tensor or capillary broadening are used to obtain γ_{IN}, the statistical errors will likely obscure any finite size dependence.

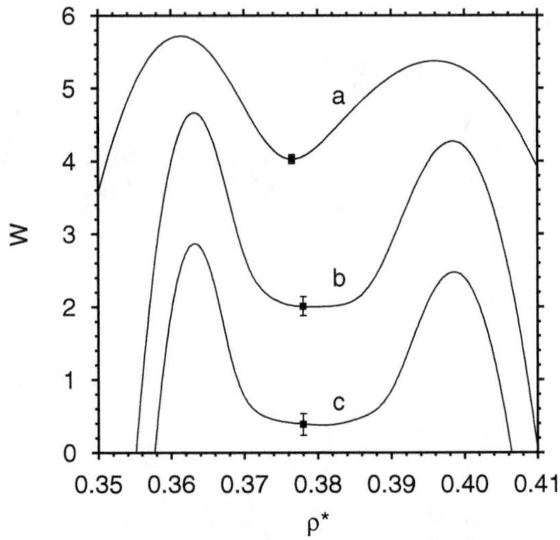

Fig. 21.8. Coexistence distributions $W = k_B T \ln P(N)$ of soft spherocylinders with $L/D = 10$ and $\varepsilon = 2$ for various system sizes. In each of the above distributions, the lateral box dimension was fixed at $L_x = L_y = 2.3\,L$, while the perpendicular dimension was varied: (**a**) $L_z = 2.3\,L$; (**b**) $L_z = 10.35\,L$; (**c**) $L_Z = 13.8\,L$. The corresponding free energy barriers ΔF are: (**a**) 1.52 ± 0.05; (**b**) 2.47 ± 0.13; (**c**) 2.29 ± 0.15, in units of $k_B T$. The error bars indicate the magnitude of the scatter in ΔF for a number of independent measurements

Table 21.1. Bulk properties of the coexisting isotropic and nematic phase in a system of soft spherocylinders with $\varepsilon = 2$ and $L/D = 10$ and 15. Listed are the reduced density $\rho^\star$ of the isotropic and the nematic phase, the normalized number density $\rho L D^2$ and the interfacial tension γ_{IN}, expressed in two types units to facilitate the comparison to other work

L/D	isotropic phase		nematic phase		interfacial tension γ_{IN}	
	$\rho^\star$	$\rho L D^2$	$\rho^\star$	$\rho L D^2$	$\frac{k_B T}{D^2}$	$\frac{k_B T}{LD}$
10	0.363	0.388	0.397	0.424	0.0022 ± 0.0003	0.033
15	0.244	0.267	0.280	0.307	0.0053 ± 0.0001	0.080

21.5 Discussion

It is clear from the phase diagram of Fig. 21.6 that the Onsager limit is not recovered until for very large rod elongation, exceeding at least $L/D = 40$. As a result, our estimates for the interfacial tension differ profoundly from Onsager predictions. Typically, γ_{IN} in our simulations is four times lower than Onsager estimates. Note that our simulations also show that γ_{IN} increases with L/D, towards the Onsager result, so there seems to be qualitative agreement. However, to properly access the Onsager regime, additional simulations for large elongation L/D are required. Unfortunately, as indicated by the scatter in the data of Fig. 21.6, and also in [12], such simulations are very complicated. It is questionable if present simulation techniques are sufficiently powerful to extract $\gamma_{\rm IN}$ with any meaningful accuracy in the Onsager regime.

As mentioned in the introduction, computer simulations of soft ellipsoids with $\kappa = 15$ yield interfacial tensions of $\gamma_{\rm IN} = 0.011 \pm 0.004\ k_BT/B^2$ and $\gamma_{\rm IN} = 0.016 \pm 0.002\ k_BT/B^2$ [18,19]. For $L/D = 15$, our result for soft spherocylinders is considerably lower. Obviously, spherocylinders are not ellipsoids, and this may well be the source of the discrepancy. Note also that the shape of the potential used by us is different from that of [18,19].

In summary, we have performed grand canonical Monte Carlo simulations of the IN transition in a system of soft spherocylinders. By measuring the grand canonical order parameter distribution, the coexistence densities as well as the interfacial tension were obtained. In agreement with theoretical expectations and other simulations, ultra-low values for the interfacial tension $\gamma_{\rm IN}$ are found. Our results confirm that for short rods, the interfacial tension, as well as the coexistence densities, are considerably lower than the Onsager predictions. This demonstrates the need for improved theory to describe the limit of shorter rods, which is required if the connection to experiments is ever to be made.

Acknowledgement

We are grateful to the Deutsche Forschungsgemeinschaft (DFG) for support (TR6/A5) and to K. Binder, M. Müller, P. van der Schoot, and R. van Roij for stimulating discussions. We also thank G. T. Barkema for suggesting some of the numerical optimizations used in this work. T.S was supported by the Emmy Noether program of the DFG.

References

1. L. Onsager: Ann. N. Y. Acad. Sci. **51**, 627 (1949)
2. M. Doi, N. Kuzuu: Appl. Polym. Symp. **41**, 65 (1985)
3. W.E. McMullen: Phys. Rev. A **38**, 6384 (1988)
4. Z.Y. Chen, J. Noolandi: Phys. Rev. A **45**, 2389 (1992)
5. Z.Y. Chen: Phys. Rev. E **47**, 3765 (1993)
6. D.L. Koch, O.G. Harlen: Macromolecules **32**, 219 (1999)
7. K. Shundyak, R. van Roij: J. Phys.: Condens. Matter **13**, 4789 (2001)
8. E. Velasco, L. Mederos, D.E. Sullivan: Phys. Rev. E **66**, 021708 (2002)
9. P. van der Schoot: J. Phys. Chem. B **103**, 8804 (1999)
10. W. Chen, D.G. Gray: Langmuir **18**, 663 (2002)
11. M. Dijkstra, R. van Roij, R. Evans: Phys. Rev. E **63**, 051703 (2001)
12. P. Bolhuis, D. Frenkel: J. Chem. Phys. **106**, 666 (1997)
13. M.A. Bates, C. Zannoni: Chem. Phys. Lett. **280**, 40 (1997)
14. M.A. Bates, C. Zannoni: Chem. Phys. Lett. **288**, 209 (1998)

15. M.P. Allen: J. Chem. Phys. **112**, 5447 (2000)
16. M.P. Allen: Chem. Phys. Lett. **331**, 513 (2000)
17. M.S. Al-Barwani, M.P. Allen: Phys. Rev. E **62**, 6706 (2000)
18. A.J. McDonald, M.P. Allen, F. Schmid: Phys. Rev. E **63**, 010701 (2000)
19. N. Akino, F. Schmid, M.P. Allen: Phys. Rev. E **63**, 041706 (2001)
20. Q. Yan, J.J. de Pablo, J. Chem. Phys **113**, 1276 (2000)
21. P. Virnau, M. Müller: J. Chem. Phys. **120**, 10925 (2004)
22. J. Potoff, A. Panagiotopoulos: J. Chem. Phys. **112**, 6411 (2000)
23. W. Góźdź: J. Chem. Phys. **119**, 3309 (2003)
24. P. Virnau, M. Müller, L.G. MacDowell, K. Binder: J. Chem. Phys. **121**, 2169 (2004)
25. R.L.C. Vink, J. Horbach: J. Chem. Phys. **121**, 3253 (2004)
26. K. Binder: Phys. Rev. A **25**, 1699 (1982)
27. D.P. Landau, K. Binder: *A Guide to Monte Carlo Simulations in Statistical Physics* (Cambridge University Press, Cambridge 2000)
28. A.D. Bruce, N.B. Wilding: Phys. Rev. Lett. **68**, 193 (1992)
29. Y.C. Kim, M.E. Fisher, E. Luijten: Phys. Rev. Lett. **91**, 065701 (2003)
30. D. Frenkel, B. Smit: *Understanding Molecular Simulation*(Academic Press, San Diego 2001)
31. M. Müller, N.B. Wilding: Phys. Rev. E **51**, 2079 (1995)
32. R. Vink, T. Schilling: cond-mat/0502444 (2005)
33. N.B. Wilding: Am. J. Phys. **69**, 10 (2001)

22

Langevin Dynamics Study of Polymer Translocation through a Nanopore

L. Guo and E. Luijten

University of Illinois at Urbana-Champaign, Urbana, Illinois 61801, U.S.A.

Abstract. We study the translocation of a polymer through a nanopore by means of Langevin dynamics simulations. For different driving forces F_{driving} we find different power-law scaling behavior for the polymer translocation time τ_{trans} as a function of the degree of polymerization N. If F_{driving} is small the polymer stays in or near equilibrium during translocation and the observed behavior is consistent with predictions from equilibrium theory. For large F_{driving} the polymer is driven far out of equilibrium. In this case, we observe a linear relation between τ_{trans} and N, in apparent contradiction with a theoretically suggested lower bound.

22.1 Introduction

The study of polymer translocation through a nanopore has attracted much attention in recent years. Understanding translocation processes not only advances our knowledge of the ubiquitous biological process of the transport of DNA and RNA between nucleus and cytosol in cells, but also can be relevant for the design of a DNA-sequencing device [1].

Polymer translocation takes place in a confined geometry. Both the pore and the pore-containing membrane restrict the motion of the polymer, resulting in properties that differ from polymers in bulk solution. In experiments carried out to study polymer translocation [2–4], a driving force F_{driving} (e.g., an electrical force for DNA) is usually applied across the pore to facilitate translocation. The magnitude of F_{driving} affects the translocation behavior. A directly measurable quantity in experiments is the translocation time τ_{trans}. Its dependence on the degree of polymerization N has been used to characterize the translocation dynamics. A linear scaling $\tau_{\text{trans}} \propto N$ has been observed for equilibrated systems [3, 4], where the characteristic relaxation time of the polymer τ_{relax} is much smaller than τ_{trans}. Instead, if $\tau_{\text{relax}} \gg \tau_{\text{trans}}$, a non-linear power-law $\tau_{\text{trans}} \propto N^{1.27\pm0.03}$ has been observed [2].

Both equilibrium and non-equilibrium translocation have been modeled theoretically [2, 5–8]. For the equilibrated system, polymer translocation can

Springer Proceedings in Physics, Volume 105
Computer Simulation Studies in Condensed-Matter Physics XVIII
Eds.: D.P. Landau, S.P. Lewis and H.-B. Schüttler

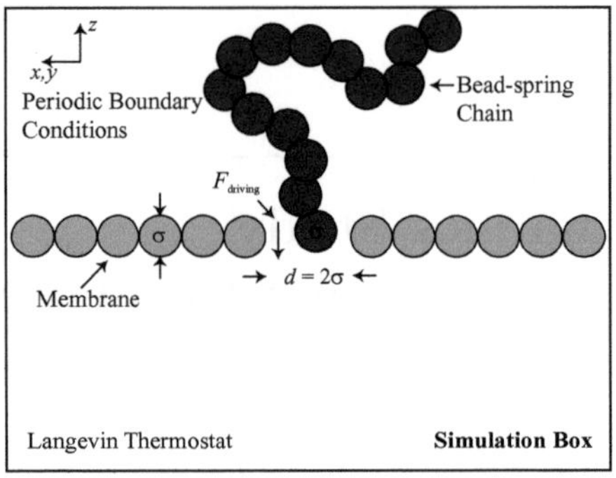

Fig. 22.1. Setup of the simulation cell. Periodic boundary conditions are applied in all spatial directions. The membrane is composed of one layer of fixed particles. The polymer is modeled as a bead–spring chain with a harmonic spring connecting neighboring monomers and a shifted-truncated Lennard-Jones interaction between monomers. A constant F_{driving} inside the pore mimics the electrical field employed in experiments

be treated as a stochastic diffusion process of a particle across an energy barrier [5, 6] and a linear relation $\tau_{\text{trans}} \propto N$ has been predicted [6] under the condition $N|\Delta\mu| \gg k_{\text{B}}T$, where $\Delta\mu \propto F_{\text{driving}}$ is the chemical potential difference between the *cis* and the *trans* side of the membrane for a single monomer and $k_{\text{B}}T$ is the thermal energy. For the non-equilibrated system, a scaling model [2] has been proposed that relies on the assumption that F_{driving} is balanced by the hydrodynamic drag force F_{drag} on the polymer [2]. This leads to the prediction $\tau \propto N^{2\nu}$ for the non-free-draining case [2] and $\tau \propto N^{1+\nu}$ for the free-draining case [2, 8], with $\nu = 0.588$ in a good solvent.

To test the assumptions of the different theories, detailed knowledge of the chain structure and the forces on the polymer is required, which cannot be obtained directly from current experiments. Computer simulation has emerged as a potential tool for this task. In this paper, we apply Langevin dynamics simulations to study the translocation of a polymer. The results are compared to experiments and used to test the theoretical predictions.

22.2 Simulation Model and Techniques

We employ a modified version of the simulation package LAMMPS [9]. The simulation box contains a membrane that is oriented perpendicular to the z-axis and is composed of one layer of fixed particles of diameter σ (Fig. 22.1). The pore in the membrane has a diameter 2σ. The polymer is represented by a bead–spring model of N monomers ($20 \leq N \leq 320$) with a harmonic bond potential $U_{\text{bond}} = K(r - r_0)^2$, where $K = 200 k_{\text{B}}T/\sigma^2$ and $r_0 = \sqrt[6]{2}\sigma$. Monomer–monomer and monomer–membrane pairs interact via a shifted-truncated Lennard–Jones interaction,

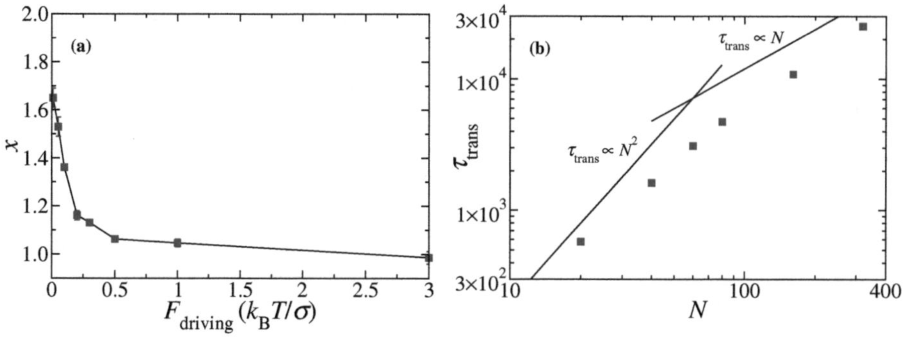

Fig. 22.2. (**a**) The exponent x for the scaling relation $\tau_{\mathrm{trans}} \propto N^x$ as a function of F_{driving}. (**b**) The scaling of τ_{trans} with chain length N for $F_{\mathrm{driving}} = 0.05 k_{\mathrm{B}}T/\sigma$, on a log–log scale. A crossover is observed between the asymptotic regimes (*solid lines*) predicted by equilibrium theory (22.2a)and (22.2b)

$$U_{\mathrm{LJ}} = 4k_{\mathrm{B}}T \left\{ \left[\left(\frac{\sigma}{r}\right)^{12} - \left(\frac{\sigma}{r}\right)^{6} \right] + \frac{1}{4} \right\} , \quad r < r_0 . \tag{22.1}$$

Periodic boundary conditions are applied in all spatial directions. To facilitate the translocation, a constant driving force F_{driving} is applied to monomers *residing in the pore*. Since our study is restricted to the translocation process itself, the first monomer is placed inside the pore at the start of the simulation and is not allowed to diffuse into the *cis* side during the simulation. Temperature is controlled by means of a Langevin thermostat with a damping coefficient $\xi = 0.1\tau_0^{-1}$ (τ_0 is the reduced time unit). The integration uses the velocity Verlet algorithm. In order to probe the translocation event with high spatial and temporal resolution, the time step is fixed at $0.005\tau_0$, even though a larger time step would have been possible for these systems.

22.3 Results and Discussion

In our simulations, we study the effect of the magnitude of the driving force F_{driving} for a range of chain lengths N. For each F_{driving} between $0.01k_{\mathrm{B}}T/\sigma$ and $3.0k_{\mathrm{B}}T/\sigma$, the calculated translocation time τ_{trans} follows a power-law $\tau_{\mathrm{trans}} \propto N^{x}$[1]. Figure 22.2a displays the scaling exponent x as a function of F_{driving}. While experimentally only two different exponents have been observed [2–4], we find a continuous variation of x as a function of the external driving force. At the strongest force $F_{\mathrm{driving}} = 3.0k_{\mathrm{B}}T/\sigma$, a linear scaling behavior is found, $x = 0.99 \pm 0.03$. This is in apparent agreement with the experimental findings for equilibrated systems [3,4], but it must be noted that

[1] Since a different crossover takes place if N becomes large (see below), we only utilize $N \leq 80$ here.

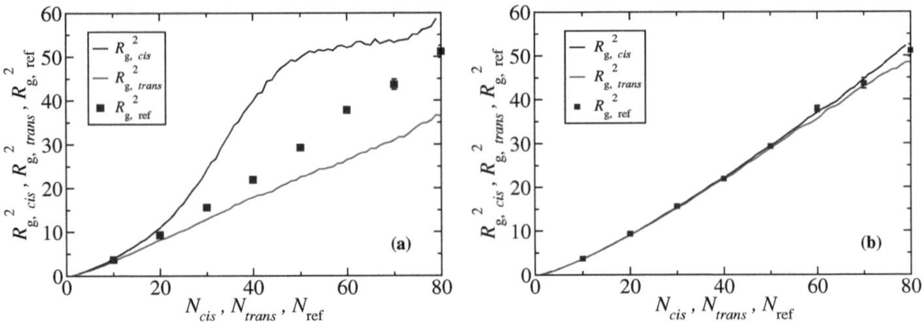

Fig. 22.3. (**a**) The squared radii-of-gyration $R^2_{g,\mathrm{cis}}$ and $R^2_{g,\mathrm{trans}}$ for the *cis* and *trans* parts of the polymer with $N = 80$ and $F_{\mathrm{driving}} = 3.0k_{\mathrm{B}}T/\sigma$, as a function of N_{cis} and N_{trans}, respectively. $R^2_{g,\mathrm{ref}}$ (*squares*) represents the equilibrium values for a chain with one end fixed at the surface of the membrane. (**b**) The same figure with $F_{\mathrm{driving}} = 0.05k_{\mathrm{B}}T/\sigma$

our system is driven out of equilibrium for a driving force of this magnitude. Indeed, for $N = 80$ we find $\tau_{\mathrm{relax}} \simeq 110\tau_0$[2], much larger than the translocation time per monomer $\tau_{\mathrm{trans}}/N \simeq 1.39\tau_0$. The variation of the chain structure during translocation confirms the non-equilibrium character of the system. The squared radii-of-gyration $R^2_{g,\mathrm{cis}}$ and $R^2_{g,\mathrm{trans}}$ for the *cis* and *trans* parts of the polymer, respectively, are shown in Fig. 22.3a as a function of the corresponding number of monomers N_{cis} and N_{trans} on each side of the membrane.

The square symbols in Fig. 22.3a ($R^2_{g,\mathrm{ref}}$) represent the corresponding equilibrium values for a chain of N_{ref} monomers with one end anchored to the surface of the membrane. The deviations from these reference values clearly indicate that the translocating chain has a highly non-equilibrated structure, in which the *cis* part is elongated and the *trans* part is contracted. For the free-draining system out of equilibrium, $\tau_{\mathrm{trans}} \propto N^{1+\nu}$ is predicted [2,8], which is at variance with our result. A basic assumption in [2] is that F_{driving} is balanced by F_{drag}. Our simulations suggest that this assumption is incorrect if F_{driving} is large. In this case, the *cis* part of the polymer near the membrane experiences a larger repulsive force from the surface of the membrane than its *trans* counterpart. Correspondingly, there is a net surface force F_{surf} acting on the polymer. Indeed, for $N = 80$ we find $F_{\mathrm{surf}} \approx -2.4k_{\mathrm{B}}T/\sigma$, where the minus sign indicates that the force is oriented opposite to the translocation direction. This non-negligible F_{surf} may account for the discrepancy between our result and the theoretical prediction [2]: For a large driving force, the membrane acts not only as an entropic barrier but also as an effective drag force.

For the equilibrium system, the theory [6] predicts two different regimes for the scaling relation between τ_{trans} and N:

[2] τ_{relax} is estimated from the exponential decay of the autocorrelation function $< \boldsymbol{P}(t) \cdot \boldsymbol{P}(0) >$ for the chain end-to-end vector $\boldsymbol{P}(t) \equiv \boldsymbol{R}_N(t) - \mathbf{R}_1(t)$.

$$\tau \propto N \quad \text{if} \quad N\Delta\mu \gg k_{\mathrm{B}}T\,, \tag{22.2a}$$

$$\tau \propto N^2 \quad \text{if} \quad N\Delta\mu = 0\,. \tag{22.2b}$$

The linear scaling relation (22.2a) is consistent with the experimental observations [3,4], while the quadratic dependence (22.2b) has not been tested yet. To test (22.2b), a small F_{driving} is required in order to satisfy the equilibrium condition $\tau_{\text{relax}} < \tau_{\text{trans}}/N$. In our system this is realized for $F_{\text{driving}} = 0.05k_{\mathrm{B}}T/\sigma$ and $N < 60$. In this range of chain lengths, we observe an effective power-law dependence with an exponent that slowly approaches 2, as shown in Fig. 22.2b. This is consistent with a gradual approach of (22.2b), as confirmed by the observation that $N\Delta\mu = NF_{\text{driving}}\sigma < 3k_{\mathrm{B}}T$ for $N < 60$. As N increases ($N \geq 80$ in Fig. 22.2b), the apparent scaling exponent decreases. Whereas the equilibrium condition is no longer fully satisfied for these chains (e.g., for $N = 80$, $\tau_{\text{relax}} = 110\tau_0 > \tau_{\text{trans}}/N = 50\tau_0$), the measured $R^2_{g,\text{cis}}$ and $R^2_{g,mathrmtrans}$ closely follow the corresponding equilibrium values $R^2_{g,\text{ref}}$ (Fig. 22.3b), indicating that the system stays near equilibrium. Therefore, we conclude that the decrease of the scaling exponent is indeed a direct consequence of the increase of $N\Delta\mu$, which leads to a crossover from (22.2b) to (22.2a).

22.4 Conclusions

We have studied the translocation of a polymer through a nanopore by means of Langevin dynamics simulations, as a function of external driving force F_{driving} and degree of polymerization N. A range of scaling exponents is found, which depend on F_{driving}. For small F_{driving} the polymer remains in or near equilibrium during translocation. The observed power-law dependence of τ_{trans} on N is consistent with the predictions of the equilibrium theory [6]. For large F_{driving} the polymer is driven out of equilibrium. A linear scaling behavior between τ_{trans} and N is recovered for such a system, where the repulsive force exerted by the membrane surface becomes non-negligible and acts as an effective drag force.

Acknowledgment

This work is supported by the American Chemical Society Petroleum Research Fund under Grant No. 38543-G7 and by the National Science Foundation through an ITR grant (DMR-03-25939) via the Materials Computation Center at the University of Illinois at Urbana-Champaign. The authors thank Intel for a generous equipment donation.

References

1. D.W. Deamer, M. Akeson: Trends in Biotech. **18**, 147 (2000)
2. A.J. Storm, C. Storm, J. Chen, H. Zandbergen, J.-F. Joanny, C. Dekker: Nano Letters (2005)
3. J.J. Kasianowicz, E. Brandin, D. Branton, D.W. Deamer: Proc. Natl. Acad. Sci. U.S.A. **93**, 13770 (1996)
4. A. Meller, L. Nivon, D. Branton: Phys. Rev. Lett. **86**, 3435 (2001)
5. W. Sung, P.J. Park: Phys. Rev. Lett. **77**, 783 (1996)
6. M. Muthukumar: J. Chem. Phys. **111**, 10371 (1999)
7. D.K. Lubensky, D.R. Nelson: Biophys. J. **77**, 1824 (1999)
8. Y. Kantor, M. Kardar: Phys. Rev. E **69**, 021806 (2004)
9. S.J. Plimpton: J. Comp. Phys. **117**, 1 (1995)

23

Invasive Allele Spread under Preemptive Competition

J.A. Yasi[1], G. Korniss[1], and T. Caraco[2]

[1] Department of Physics, Applied Physics, and Astronomy,
Rensselaer Polytechnic Institute, 110 8th Street, Troy, NY 12180-3590, USA
E-mail: yasij@rpi.edu, korniss@rpi.edu

[2] Department of Biological Sciences,
University at Albany, Albany NY 12222, USA
E-mail: caraco@albany.edu

Abstract. We study a discrete spatial model for invasive allele spread in which two alleles compete preemptively, initially only the "residents" (weaker competitors) being present. We find that the spread of the advantageous mutation is well described by homogeneous nucleation; in particular, in large systems the time-dependent global density of the resident allele is well approximated by Avrami's law.

23.1 Introduction and Model

The spatial and temporal characteristics of the spread of an advantageous mutation are fundamental questions in population dynamics. Fisher [1] and Kolmogorov et al. [2] first addressed these questions using the framework of a simple reaction-diffusion equation [3]. That work and many others focused on the velocity of the propagating front, which exists initially and separates the two spatial regions occupied separately by the two alleles. Both continuum and discrete spatial models have successfully tackled various aspects of these problems [3, 4].

In this work we investigate how the advantageous allele emerges from "scratch"; i.e., initially the region is fully dominated by the resident allele and the advantageous allele is introduced by rare mutations. While the mutant allele has an individual-level advantage over the original one, the low probability of mutations, combined with a discrete spatial dynamics, can prevent the spread of the mutant for long times. Here we consider a model where the original "resident" and the competitively superior "invasive" allele compete for a common limiting resource preemptively [5–8].

The details of our model are as follows. We consider an $L \times L$ lattice with periodic boundary conditions. Each site can be empty or occupied by a *single* allele (either a resident or an invader). A lattice site represents the minimal

Springer Proceedings in Physics, Volume 105
Computer Simulation Studies in Condensed-Matter Physics XVIII
Eds.: D.P. Landau, S.P. Lewis and H.-B. Schüttler

level of locally available resource required to sustain an individual organism, hence the "excluded volume" constraint. We introduce the local occupation numbers at site $\boldsymbol{x}$, $n_i(\boldsymbol{x}) = 0, 1$, $i = 1, 2$, representing the number of resident and invader alleles, respectively. By virtue of the excluded volume constraint, $n_1(\boldsymbol{x})n_2(\boldsymbol{x}) = 0$. New individuals arise through local clonal propagation only. That is, an individual occupying site $\boldsymbol{x}$ may reproduce if one or more neighboring sites are empty (here we consider nearest neighbor colonization only). Competition for resources, hence space, is preemptive, therefore, an occupied site cannot be colonized by either allele until the current occupant's mortality leaves that site empty. Each individual may mutate and so carry the alternate allele; mutation is a two-way, recurrent process.

We performed dynamic Monte Carlo simulations to study the above model. Our time unit is one Monte Carlo step per site (MCSS) during which L^2tes are chosen randomly. If a site is empty, it may be colonized by individuals of allele i occupying neighboring sites, at the rate $\alpha_i \eta_i(\boldsymbol{x})$; α_i is the individual-level colonization rate and $\eta_i(\boldsymbol{x}) = (1/4)\sum_{\boldsymbol{x}' \varepsilon \mathrm{nn}(\boldsymbol{x})} n_i(\boldsymbol{x}')$ is the density of allele i around site $\boldsymbol{x}$ [nn($\boldsymbol{x}$) is the set of nearest neighbors of site $\boldsymbol{x}$]. If a site is occupied by an individual, it can die at rate μ (regardless of the allele) or mutate to the other allele at rate ϕ. We can summarize the local transition rules for an arbitrary site $\mathbf{x}$ as

$$0 \stackrel{\alpha_1\eta_1(\mathbf{x})}{\longrightarrow} 1\,, \quad 0 \stackrel{\alpha_2\eta_2(\mathbf{x})}{\longrightarrow} 2\,, \quad 1 \stackrel{\mu}{\longrightarrow} 0\,, \quad 2 \stackrel{\mu}{\longrightarrow} 0, \quad 1 \stackrel{\phi}{\longleftrightarrow} 2\,, \tag{23.1}$$

where $0, 1, 2$ indicates whether the site is empty, occupied by an individual with the resident, or an individual with the invader allele, respectively.

In the simulations, we initialized the system fully occupied by the resident allele ($n_1(\boldsymbol{x}) = 1$ for all $\boldsymbol{x}$). We are interested in the parameter region where $\phi \ll \mu < \alpha_1 < \alpha_2$, so that mutation is a rare process, but the invader allele has a reproductive-effort advantage. Then, due to mortality, the system quickly relaxes (much too fast for mutation to play a role) to the "quasi-equilibrium" state where the resident's population is balanced by its own clonal propagation and mortality rates (in the near absence of invaders). Throughout the simulations, we track the time-dependent global densities of the two alleles, $\rho_i(t) = (1/L^2)\sum_{\boldsymbol{x}} n_i(\boldsymbol{x}, t)$. We define the lifetime τ of the resident allele as the first passage time of $\rho_1(t)$ to one-half of its quasi-equilibrium value ρ_1^*.

23.2 Single-Cluster and Multi-Cluster Spread

As a result of the rare mutations, individuals with the invasive allele occasionally appear in the population. An invader lacking access to nearby resources may die without propagating. If a site opens in the local neighborhood (resource becomes available), the invader may colonize it. However, the empty site is likely surrounded by more than one resident. The resident's greater local density can compensate for its lower individual-level colonization rate, so

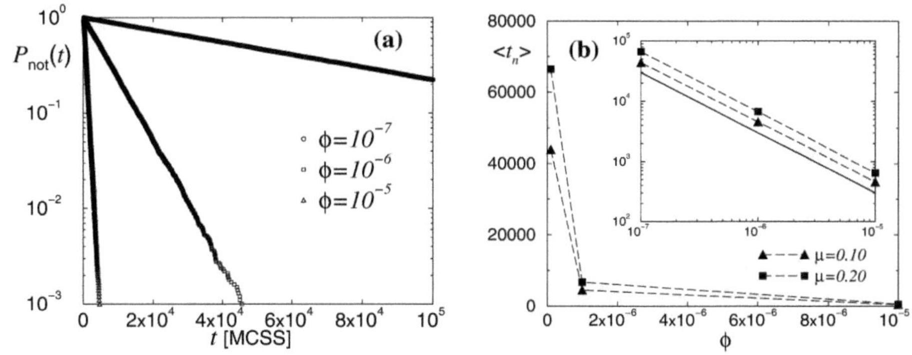

Fig. 23.1. (**a**) Cumulative probability distributions $P_{\text{not}}(t)$ for $L = 32$, $\alpha_1 = 0.50$, $\alpha_2 = 0.70$, and $\mu = 0.20$ for three different values of the mutation rate ϕ (in increasing order from the top). (**b**) Average nucleation time (in units of MCSS) vs. the mutation rate for two different values of μ [α_1, α_2, and L are the same as for (a)]. The inset shows the same on log-log scales. The straight solid line corresponds to a slope -1, indicating $\langle t_n \rangle \sim \phi^{-1}$ in the single-cluster regime

the resident has the better chance of colonizing an empty site. Consequently, one expects small clusters of the invading allele to shrink and disappear. Residents, although weaker competitors, can prevail for some time, since preemptive competition imposes a strong constraint on the growth of the invaders. Individuals with the advantageous allele can succeed only if they generate a cluster large enough that it statistically tends to grow at its periphery.

Snapshots of configurations and preliminary studies [9] confirm the existence of a critical cluster size, beyond which the spread of the invading allele becomes statistically favorable. Further, they also show strongly clustered growth of the invading allele. For a given set of parameters, there exists a length scale R_o, the typical spatial separation of invading clusters; for $L \ll R_o$ the invasion almost always occurs through the spread of a single invading cluster [single-cluster (SC) invasion], while for $L \gg R_o$ the invasion is the result of many invading clusters [multi-cluster (MC) invasion]. Conversely, fixing the linear system size L and other parameters (except the mutation rate) there is a characteristic value of ϕ (now controlling R_o), such that for sufficiently low values of ϕ, MC invasion by the advantageous allele crosses over to the SC pattern.

The above picture suggests that we can apply the framework of homogeneous nucleation and growth [10–12] to describe the spatial and temporal characteristics of the spread of the invasive allele. This framework has successfully described analogous dynamic phenomena in ferromagnetic [13–15] and ferroelectric materials [16,17], flame propagation in slow combustion [18], chemical reactions [19], and other ecological systems [9,20,21]. While local mutation is a Poisson process, lacking a Hamiltonian or an effective free energy for the model, it is not known *a priori* whether the nucleation of a "supercritical"

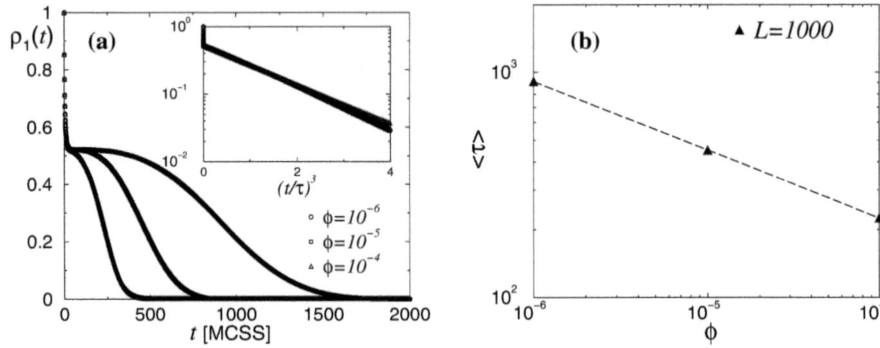

Fig. 23.2. (**a**) Time-series of the global density of the resident allele $\rho_1(t)$ for $L = 1,000$, $\alpha_1 = 0.50$, $\alpha_2 = 0.70$, and $\mu = 0.20$, for three different values of the mutation rate ϕ (in increasing order from the top). The solid curves represent Avramis's law, (23.2). The inset shows the $\rho_1(t)$ vs. t^3 on log-linear scales. (**b**) Average lifetime (in units of MCSS) vs. the mutation rate on log-log scales for the same parameter values. The straight dashed line is the best fit power-law indicating $\langle\tau\rangle \sim \phi^{-0.304}$

cluster will also be Poisson. To this end, in the SC regime, we constructed cumulative probability distributions for the lifetime of the resident allele $P_{\text{not}}(t)$, i.e., the probability that the global density of the resident has not crossed below $\rho_1^*/2$ by time t. We found that these distributions are indeed exponentials (indicating that the nucleation of a successful invading cluster is a Poisson process): $P_{\text{not}}(t) = 1$ for $t \leq t_g$ and $P_{\text{not}}(t) = \exp[-(t - t_g)/\langle t_n\rangle]$ for $t > t_g$. Here $\langle t_n\rangle$ is the average nucleation time and t_g is the close-to-deterministic growth time until the advantageous mutation dominates half the system. We show results for a fixed (sufficiently small) system size for three mutation rates in Fig. 23.1a. From the slopes of the exponentials we obtained the average nucleation times (Fig. 23.1b), hence the ϕ-dependence of the nucleation rate per unit volume $I(\phi)$. Since $\langle t_n\rangle = [L^2 I(\phi)]^{-1}$, we have $I(\phi) \sim \langle t_n\rangle^{-1} \sim \phi$. In the SC regime, the invasive spread is inherently stochastic; it is initiated and completed by the first randomly nucleated successful cluster of the advantageous allele. For very low values of ϕ, the lifetime is dominated by the very large average nucleation times, hence $\langle\tau\rangle = \langle t_n\rangle + t_g \approx \langle t_n\rangle \sim \phi^{-1}$.

In the MC regime the invasion processes becomes self-averaging and the global densities approach deterministic functions in the limit of $L \to \infty$. At the same time, $\langle\tau\rangle$ approaches a system-size independent limit. For large systems we applied the KJMA theory [10–12] (or Avrami's law) to predict the density of the resident alelle,

$$\rho_1(t) \simeq \rho_1^* \, e^{-\ln(2)(t/\langle\tau\rangle)^3} \,. \tag{23.2}$$

Our results in Fig. 23.2a show that it is, indeed, a very good approximation. Assuming that the spreading velocity of the invading clusters is constant,

KJMA theory predicts that $\langle \tau \rangle \sim [I(\phi)]^{-1/3} \sim \phi^{-1/3}$ in the MC regime. The measured exponent, −0.304, is not too far off (Fig. 23.2b), but it indicates that the assumption of a constant spreading velocity (possibly as a result of the nontrivial surface properties of the clusters) may break down.

23.3 Summary and Outlook

We studied the spread of an advantageous mutant in a two-allele population where rare mutations introduce the favored allele. We found that nucleation theory, in particular Avrami's law, describes this phenomenon very well. Systematic studies of the critical cluster size and the cluster-size dependence of the spreading velocity are under way. The structure of the spreading clusters, in particular, the roughness of their surface, is expected to play an important role in the latter.

Acknowledgement

This research is supported in part by the US NSF through Grant Nos. DMR-0113049, DMR-0426488, DEB-0342689 and by the Research Corporation through Grant No. RI0761.

References

1. R.A. Fisher: Annals of Eugenics **7**, 355 (1937)
2. A.N. Kolmogorov, I. Petrovsky, N. Piskounov: Moscow Univ. Bull. Math. **1**, 1 (1937)
3. J.D. Murray: *Mathematical Biology I and II.* 3rd edn. (Springer, Berlin Heidelberg New York 2003)
4. D. ben-Avraham, S. Havlin: *Diffusion and Reactions in Fractals and Disordered Systems.* (Cambridge University Press, Cambridge, UK 2000)
5. J.B. Shurin, P. Amarasekare, J.M. Chase, R.D. Holt, M.F. Hoopes, M.A. Leibold: J. Theor. Biol. **227**, 359 (2004)
6. P. Amarasekare: Ecol. Lett. **6**, 1109 (2003)
7. D.W. Yu, H.B. Wilson: Am. Nat. **158**, 49 (2001)
8. D.E. Taneyhill: Ecol. Monogr. **70**, 495 (2000)
9. L. O'Malley, A. Allstadt, G. Korniss, T. Caraco, in *Fluctuations and Noise in Biological, Biophysical, and Biomedical Systems III.* Ed. by N.G. Stocks, D. Abbott, and R.P. Morse, Proceedings of SPIE Vol. 5841 (SPIE, Bellingham, WA, 2005) in press
10. A.N. Kolmogorov: Bull. Acad. Sci. USSR, Phys. Ser. **1**, 355 (1937)
11. W. A. Johnson and R. F. Mehl: Trans. Am. Inst. Mining and Metallurgical Engineers **135**, 416 (1939)
12. M. Avrami: J. Chem. Phys. **7**, 1103 (1939); J. Chem. Phys. **8**, 212 (1940); J. Chem. Phys. **9**, 177 (1941)

13. P.A. Rikvold, H. Tomita, S. Miyashita, S.W. Sides: Phys. Rev. E **49**, 5080 (1994)
14. H.L. Richards, S.W. Sides, M.A. Novotny, P.A. Rikvold: J. Magnet. Mater. **150**, 37 (1995)
15. R.A. Ramos, P.A. Rikvold, M.A. Novotny: Phys. Rev. B **59**, 9053 (1999)
16. Y. Ishibashi, Y. Takagi: J. Phys. Soc. Jpn. **31**, 506 (1971)
17. H.M. Duiker, P.D. Beale: Phys. Rev. B **41**, 490 (1990)
18. M. Karttunen, N. Provatas, T. Ala-Nissila, M. Grant: J. Stat. Phys. **90**, 1401 (1998)
19. E. Machado, G. M. Buendia, P. A. Rikvold: Phys. Rev. E (2005) in press, arXiv: cond-mat/0411335
20. G. Korniss, T. Caraco: J. Theor. Biol. **233**, 137 (2005)
21. A. Gandhi, S. Levin, S. Orszag: J. Theor. Biol. **200**, 121 (1999)

List of Contributors

SPRINGER PROCEEDINGS IN PHYSICS

60 **The Physics and Chemistry of Oxide Superconductors**
Editors: Y. Iye and H. Yasuoka

61 **Surface X-Ray and Neutron Scattering**
Editors: H. Zabel and I.K. Robinson

62 **Surface Science**
Lectures on Basic Concepts and Applications
Editors: F.A. Ponce and M. Cardona

63 **Coherent Raman Spectroscopy**
Recent Advances
Editors: G. Marowsky and V.V. Smirnov

64 **Superconducting Devices and Their Applications**
Editors: H. Koch and H. Lübbing

65 **Present and Future of High-Energy Physics**
Editors: K.-I. Aoki and M. Kobayashi

66 **The Structure and Conformation of Amphiphilic Membranes**
Editors: R. Lipowsky, D. Richter, and K. Kremer

67 **Nonlinearity with Disorder**
Editors: F. Abdullaev, A.R. Bishop, and S. Pnevmatikos

68 **Time-Resolved Vibrational Spectroscopy V**
Editor: H. Takahashi

69 **Evolution of Dynamical Structures in Complex Systems**
Editors: R. Friedrich and A. Wunderlin

70 **Computational Approaches in Condensed-Matter Physics**
Editors: S. Miyashita, M. Imada, and H. Takayama

71 **Amorphous and Crystalline Silicon Carbide IV**
Editors: C.Y. Yang, M.M. Rahman, and G.L. Harris

72 **Computer Simulation Studies in Condensed-Matter Physics IV**
Editors: D.P. Landau, K.K. Mon, and H.-B. Schüttler

73 **Surface Science**
Principles and Applications
Editors: R.F. Howe, R.N: Lamb, and K. Wandelt

74 **Time-Resolved Vibrational Spectroscopy VI**
Editors: A. Lau, F. Siebert, and W. Werncke

75 **Computer Simulation Studies in Condensed-Matter Physics V**
Editors: D.P. Landau, K.K. Mon, and H.-B. Schüttler

76 **Computer Simulation Studies in Condensed-Matter Physics VI**
Editors: D.P. Landau, K.K. Mon, and H.-B. Schüttler

77 **Quantum Optics VI**
Editors: D.F. Walls and J.D. Harvey

78 **Computer Simulation Studies in Condensed-Matter Physics VII**
Editors: D.P. Landau, K.K. Mon, and H.-B. Schüttler

79 **Nonlinear Dynamics and Pattern Formation in Semiconductors and Devices**
Editor: F.-J. Niedernostheide

80 **Computer Simulation Studies in Condensed-Matter Physics VIII**
Editors: D.P. Landau, K.K. Mon, and H.-B. Schüttler

81 **Materials and Measurements in Molecular Electronics**
Editors: K. Kajimura and S. Kuroda

82 **Computer Simulation Studies in Condensed-Matter Physics IX**
Editors: D.P. Landau, K.K. Mon, and H.-B. Schüttler

83 **Computer Simulation Studies in Condensed-Matter Physics X**
Editors: D.P. Landau, K.K. Mon, and H.-B. Schüttler

84 **Computer Simulation Studies in Condensed-Matter Physics XI**
Editors: D.P. Landau and H.-B. Schüttler

85 **Computer Simulation Studies in Condensed-Matter Physics XII**
Editors: D.P. Landau, S.P. Lewis, and H.-B. Schüttler

86 **Computer Simulation Studies in Condensed-Matter Physics XIII**
Editors: D.P. Landau, S.P. Lewis, and H.-B. Schüttler

MIX
Papier aus verantwortungsvollen Quellen
Paper from responsible sources
FSC® C105338

If you have any concerns about our products,
you can contact us on
ProductSafety@springernature.com

In case Publisher is established outside the EU,
the EU authorized representative is:
Springer Nature Customer Service Center GmbH
Europaplatz 3, 69115 Heidelberg, Germany

Printed by Libri Plureos GmbH
in Hamburg, Germany